CONTENTS

Chapter 1
BASIC SEMICONDUCTOR PHYSICS

Structure of Matter

For many years, the atom was considered to be the smallest particle of matter. It is now known that the atom is composed of still smaller entities called electrons, protons, and neutrons. Each atom of any one element contains specific quantities of these electrical entities.

Physically, the electrons rotate around the core or nucleus of the atom, which contains the protons and neutrons. Figure 1-1 illustrates the layout of a carbon atom. A carbon atom contains six each of electrons, protons, and neutrons. Note that the six orbital electrons do not rotate at equal distances from the nucleus, but rather are restricted to two separate rings. With respect to the size of these electrons, tremendous distances exist between the electrons and the nucleus. If it were possible to magnify the atom by a factor of 10^{14}, that is, one hundred thousand billion times, the electrons would be the size of basketballs, with an orbit spacing of approximately 12 miles.

The negative electrical charge of the electron is exactly equal and opposite to the charge of the proton. The neutron has no charge. The electron is three times larger than the proton, but its mass is only .0005 that of the proton. In an electrically balanced atom, as illustrated in Fig. 1-1, there is an equal number of electrons and protons.

Gravitational, electric, magnetic, and nuclear forces all act within the atom. These forces tend to keep the electrons revolving in their orbits around the nucleus at tremendous speeds. As might be expected, the electrons located in rings close to the nucleus are tightly bound to their orbit and are extremely difficult to dislodge. The outer or so-called *valence-ring* electrons are, comparatively speaking, loosely bound to their orbit. The ease or difficulty with which electrons can be dislodged from the outer orbit determines whether a particular element is a conductor, insulator, or semiconductor.

Conductors, Insulators, Semiconductors

Conductors are materials that have a large number of loosely bound valence-ring electrons; these electrons are easily knocked out of their orbit and are then referred to as free electrons. *Insulators* are materials in which the valence-ring electrons are tightly bound to the nucleus. In between the limits of these two major categories is a third general class of materials called *semiconductors*. For example, transistor germanium, a semiconductor, has approximately one trillion times (1×10^{12}) the conductivity of glass, an insulator, but has only about one thirty-millionth (3×10^{-8}) part of the conductivity of copper, a conductor.

The heart of the transistor is a semiconductor, generally the germanium crystal. Other semiconductors such as selenium and silicon have

1

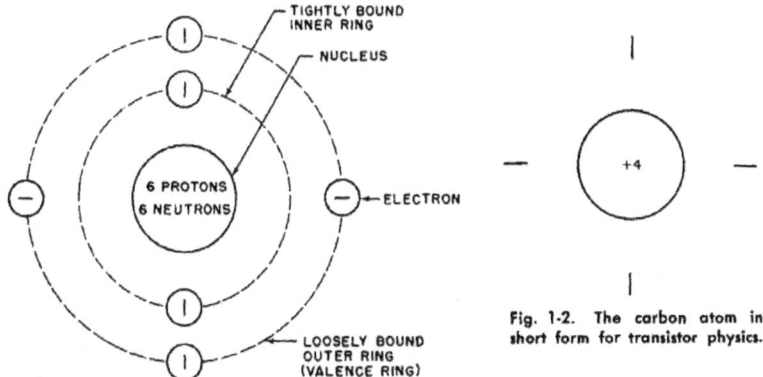

Fig. 1-1. The carbon atom.

Fig. 1-2. The carbon atom in short form for transistor physics.

been used in transistors, but germanium has proved to be the most widely applicable material. The general semiconductor principles discussed in this book apply to all elements used as transistor semiconductors.

Insofar as transistor operation is concerned, only the loosely bound orbital electrons and their associated protons are of importance. For the purposes of future discussion it is therefore convenient to picture the carbon atom in the short form illustrated in Fig. 1-2. Note that in this figure only the valence-ring electrons and their associated protons are indicated; the tightly bound inner orbit electrons and their respective protons are not shown. Thus, the carbon atom in the short form contains a nucleus with a +4 charge around which the four valence-ring electrons rotate. The short form simplifies the graphical representation of semiconductor operation, as will be seen later.

Crystal Structure

Covalent Bonds. Carbon is occasionally found in nature in a stable crystalline form, the diamond. In this form, each valence-ring electron, moving around the nucleus of a carbon atom, co-ordinates its motion with that of a corresponding valence-ring electron of a neighboring atom. Under these conditions, the electron pair forms a *covalent bond.* Equilibrium between the repulsion and attraction forces of the atoms is reached at this time, the previously loosely bound valence-ring electrons now are tightly bound to their nucleus, and cannot easily be dislodged. This effectively reduces the number of available free electrons in the crystal, and hence reduces its conductivity. Thus, carbon, generally a semiconductor, becomes an insulator in the diamond form.

The Germanium Crystal. Like the carbon atom, the germanium atom has four valence-ring electrons. Thus a short-form illustration of the germanium atom would be similar to that shown for the carbon atom in Fig. 1-2. In addition, when germanium is in crystalline form, the four valence electrons of each atom form covalent bonds, and are tightly bound to the nucleus. Figure 1-3 (A) is a short-form illustration of the structure of the germanium crystal in this perfect state. (For simplification, the atoms in this figure are shown in a two-dimensional plane rather than in the three dimensions found in nature.) Note that all co-valent bonds are complete and that no atoms or electrons are missing or misplaced. The pure germanium crystal is an insulator and is of no use in transistor work. However, pure germanium can be changed into a semiconductor by adding minute quantities of certain impurities, or by adding heat energy (phonons), or by adding light energy (photons). Any of these actions increases the number of free electrons in germanium.

If an excess free electron could be added to a pure germanium crys-tal without changing the structure of the crystal, the electron would move through the crystal as freely as an electron moves through a vacuum tube. However, when pure germanium is treated so as to become a semi-

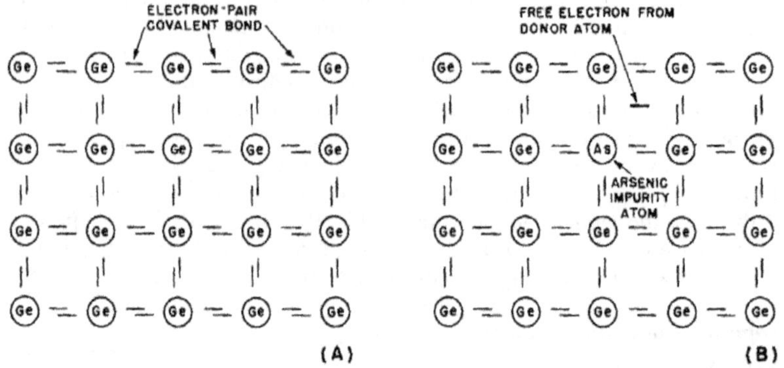

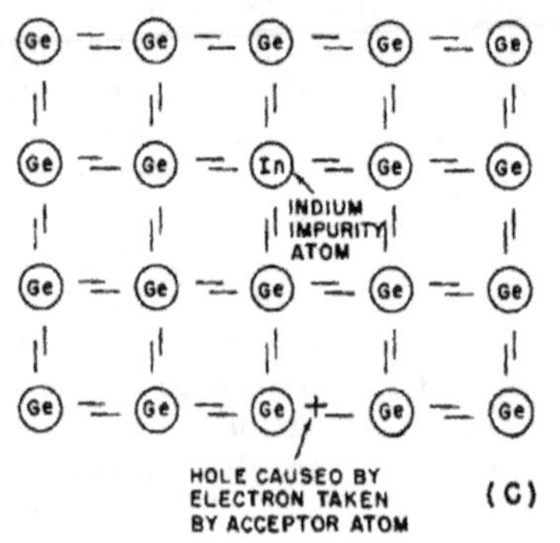

(A)

(B)

Fig. 1-3. (A) Pure germanium in crystalline form. (B). N-type germanium. (C) N-type ger-manium.

(C)

conductor, the symmetry of the crystal is destroyed. Consequently, any one excess electron moves a short distance, bounces off an imperfection, and then moves on again. This collision is similar to the collision between an electron and a gas molecule in a gas tube.

Donors — N-Type Germanium. When impurities having five electrons in the valence ring are added to germanium, each impurity atom replaces a germanium atom. Four of the impurity atom's valence electrons form covalent bonds with the valence electrons of neighboring germanium atoms. The fifth electron is free and is available as a current carrier. Pentavalent-type impurities are called *donors* because they donate electrons to the crystal transistor germanium thus formed. Such transistor crystals are referred to as *N* type because conduction is carried on by means of the *N*egatively charged electrons, contributed by the donor atoms. This action is illustrated by Fig. 1-3 (B), with arsenic acting as the pentavalent impurity.

The application of a d-c potential across the N-type crystal forces the free electrons toward the positive voltage terminal. Every time an electron flows from the crystal to the positive terminal, an electron enters the crystal through the negative voltage terminal. In this manner a continuous stream of electrons flows through the crystal as long as the battery potential remains.

Acceptors — P-Type Germanium; Holes. Figure 1-3 (C) illustrates a second method of forming transistor germanium. In this case an impurity having three valence electrons (indium) is added to the pure germanium crystal. Each such trivalent impurity atom replaces a germanium atom, and in order to complete its covalent bond with neighboring germanium atoms, the impurity atom borrows a fourth electron from any one of the other germanium groups. This destruction of a germanium covalent bond group forms a *hole. A hole is an incomplete group of covalent electrons which simulates the properties of an electron with a positive charge.* These trivalent-type impurities are called *acceptors* because they take electrons from the germanium crystal. Germanium containing acceptor impurities is called *P* type because conduction is effected by *P*ositive charges.

Connection of a battery across a P-type crystal causes the holes to move toward the negative terminal. When a hole reaches the negative terminal, an electron is emitted from this battery terminal and cancels the hole. At the same time, an electron from one of the covalent bonds enters the positive terminal, thus forming another hole in the vicinity of the positive terminal. The new hole again moves towards the negative terminal. Thus the battery causes a continuous stream of holes to flow through the crystal. Insofar as the flow of current is concerned, hole flow from the positive to the negative terminal of the crystal has the same effect as electron flow from the negative to the positive terminal.

Transistor Germanium Properties

Impurity Concentration. It is interesting to note the important role that donor and acceptor atoms play in determining the conductivity of germanium. If one impurity atom is added for every 100,000,000 germanium atoms, the conductivity increases 16 times. This concentration forms germanium suitable for transistor work. If one impurity atom for each 10,000,000 germanium atoms is added, the conductivity increases 160 times, and is too high for transistor applications.

Other types of impurities which are neither trivalent nor pentavalent may be present in the crystal. These impurities are not desirable. Although they do not affect the conductivity, they introduce imperfections in the structure, and cause degradations in the transistor characteristics. Conductivity is affected, however, by the presence of N-type impurities in P-type germanium and by P-type impurities in N-type germanium, since in either case the holes furnished by the P type will cancel the electrons furnished by the N type. If both N and P types were present in equal amounts, the germanium would act as if no impurities were present. To avoid these possibilities, the germanium is purified so that the impurity ratio is considerably less than 1 part in 100,000,000 before the desired impurity atoms are added.

Intrinsic Germanium. In those cases where the germanium is extremely pure, or where there are equal numbers of donor and acceptor atoms, the germanium is called *intrinsic*. Conduction can take place if electrons are forced out of their valence bonds by the addition of external energy to the crystal in the form of heat or light. Although the disruption of the covalent bonds by these processes creates equal numbers of electrons and holes, intrinsic conduction is invariably of the N type, because the mobility of the electrons is approximately twice as great as that of the holes.

In the case of thermal excitation, the higher the temperature, the greater the number of electrons liberated and the higher the germanium conductivity becomes. This explains why germanium has a negative temperature coefficient of resistance, i.e., the higher the temperature, the lower the resistance. Intrinsic conductivity can adversely affect impurity-type conductivity. As the temperature is increased to 80° C, the electrons produced by thermal excitation cause the conductivity of the germanium to become too high for satisfactory transistor operation.

The disruption of covalent bonds by the addition of light energy is discussed under P-N junction photocells in Chapter 2.

P-N Junctions

Potential Hills. Alone, either P- or N-type germanium is capable of bi-directional current flow. This means that reversing the battery will reverse the direction of the current flow, but will not affect the magnitude of the current. When P- and N-type germanium are joined as shown

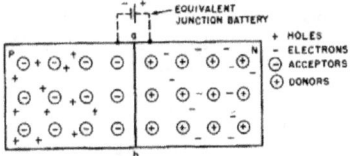

Fig. 1-4. P-N junction at equilibrium.

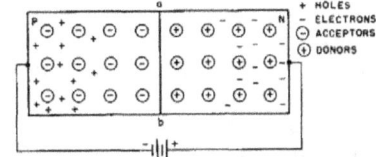

Fig. 1-5. P-N junction with reverse bias.

in Fig. 1-4, an effective rectifying device is formed. The junction, designated *ab*, is called a P-N junction. In the illustration the $+$ and $-$ signs
represent *holes* and *electrons*, respectively; and the $+$ and $-$ signs with
circles around them represent the donor and acceptor atoms, respectively.

It might appear that the holes of the P-region would diffuse into the
N-region and the electrons of the N-region would diffuse in the P-region,
eventually destroying the P-N junction. Instead, the holes and electrons
concentrate *away* from the junction. This phenomenon is caused by the
fixed position of the donor and acceptor atoms in the crystal lattice structure, as compared to the mobility of the electrons and holes. The donor
atoms repel the holes to the left in the diagram, while the acceptor atoms
repel the electrons to the right. This barrier to the flow of holes and
electrons is called a *potential hill*, and it produces the same effect as a
small battery (shown dotted in Fig. 1-4) with its negative terminal connected to the P-region and its positive terminal connected to the N-region.
To use the P-N junction as a rectifying device requires connection of
an external battery to either aid or oppose the equivalent potential hill
battery.

Reverse and Forward Bias. The connection of an external battery,
illustrated in Fig. 1-5, is an example of *reverse bias*. The negative terminal attracts holes and concentrates them further to the left, while the
positive terminal concentrates the electrons further to the right. There
is no flow across the junction, since the effect of this connection is to
increase the potential hill barrier.

Consider now the connection illustrated in Fig. 1-6 (A). This is an
example of a *forward bias* connection. The positive terminal pushes the
holes towards the N-area, while the negative terminal forces the electrons
toward the P-area. In the region around the *ab* junction, holes and electrons combine. For each combination, a covalent bond near the positive
terminal breaks down, and the liberated electron enters the positive terminal. This action creates a new hole which moves toward the N-region.
Simultaneously, an electron enters the crystal through the negative battery terminal and moves toward the P-region. The total current (I_o) flowing through the crystal is composed of electron flow (I_N) in the N-area,

hole flow (I_P) in the P-area, and a combination of the two $(I_N$ and $I_P)$ in the region near the junction. The forward bias connection, then, reduces the potential hill by a sufficient amount to allow current to flow by a combination of hole and electron carriers, as illustrated in Fig. 1-6 (B).

One may well ask, "How much battery voltage is necessary?" Offhand, since the equivalent battery potential is in the neighborhood of a few tenths of a volt, an external battery of equal value should normally be considered sufficient. Unfortunately, a large part of the battery potential

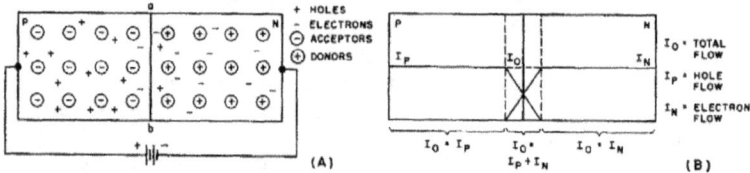

Fig. 1-6. (A) P-N junction with forward bias. (B) Carrier conduction in P-N junction.

is dropped across the resistance of the P- and N-regions before the potential hill is reached. The voltage drop in these regions is proportional to the current flow through them; as the current increases due to the reduction of the potential hill, the drop across the P- and N-regions also increases, leaving even less of the external voltage available to reduce the junction barrier potential. An external battery of approximately one to two volts is required because of these factors.

Chapter 2
TRANSISTORS AND THEIR OPERATION

In this chapter, the basic concepts concerning P- and N-type germanium are applied to an analysis of the point-contact, the junction, and certain other modified transistors. The construction, operation, gain, and impedance characteristics of typical transistors are considered.

Point-Contact Transistors

Construction and Electrode Designations. The elements and basic construction details of the point-contact transistor are shown in Fig. 2-1. This transistor consists of two electrodes (emitter and collector) which which make contact with a germanium pellet, and a third electrode (the base) which is soldered to that pellet. (It is common practice to designate the electrodes by e, c, and b—emitter, collector, and base. The practice will be followed in this book.) The entire assembly is encased in a plastic housing to avoid the contaminating effects of the atmosphere.

The pellet is usually N-type germanium, roughly .05 inch in length and .02 inch thick. The emitter and collector contacts are metallic wires, approximately .005 inch in diameter and spaced about .002 inch apart. These contacts are frequently referred to as "cat whiskers." The bend in the cat whiskers, illustrated in Fig. 2-1, is required to maintain pressure against the germanium pellet surface. The practical man will certainly ask, "Why use cat whiskers which are obviously difficult to manufacture and which produce a mechanically weak contact? Let us eliminate the cat whiskers (he goes on) and use a low-resistance soldered contact similar to that used on the base electrode." An answer to this question necessitates an analysis of the point-contact transistor. Transistor operation requires an intense electric field. If the external battery potential is made high enough to produce the required field intensity, this potential has adverse effects on the transistor. The high voltage, in the input or emitter circuit, produces a high current which burns out the transistor. In the output or collector circuit, the high voltage causes a breakdown. Thus, since the battery voltage is limited, as shown by the considerations, the use of the point-contact cat whiskers is a convenient method of obtaining the required high-intensity field. The electrical action of the points in concentrating the battery potential to produce a concentrated electric field is analagous to the increased water pressure which is obtained by decreasing the nozzle area of a garden hose.

Surface-Bound Electrons. The fundamental concepts of current flow in the point-contact transistor are illustrated in Fig. 2-2. Physicists have found that those electrons which diffuse to the surface of the germanium pellet not only lose their ability to return to the interior of the germanium but also form a skin-like covering over the surface. Because of this

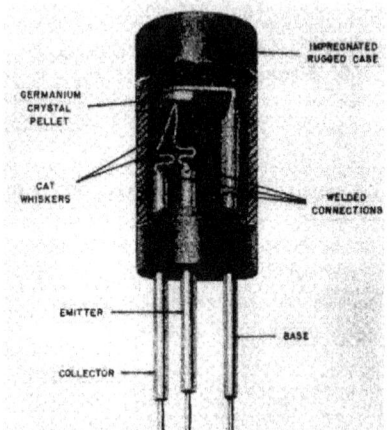

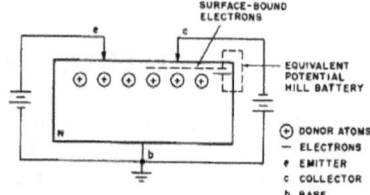

Fig. 2-1 (left). Construction of point-contact transistor. *Courtesy CBS-Hytron.*

Fig. 2-2 (right). Basic point-contact transistor operation.

phenomenon, they are called *surface-bound electrons*. For the N-type transistor illustrated, the surface-bound electrons combine with the layer of donor atoms just below to form a potential hill.

The proper battery connections for a transistor can be determined as follows: The emitter is always biased in the forward or low resistance direction. Since this is accomplished by reducing the potential hill, the positive battery terminal is connected to the emitter. Conversely, the collector is always biased in the reverse, or high-resistance, direction. Therefore, the negative battery terminal is connected to the collector in order to increase the potential hill.

Hole Injection. To understand hole and electron flow in the point-contact transistor, observe that in Fig. 2-3 the surface-bound electrons near the emitter contact are immediately removed by the positive emitter electrode. This is due to the intense emitter field which breaks down covalent bonds of atoms in the vicinity of the emitter electrode. The liberated electrons are immediately attracted to and enter the emitter terminal. These electrons are the emitter current carriers. For every electron which leaves the pellet, a hole is left behind. This creation of holes is called *hole injection,* since the effect is the same as if holes were injected into the transistor through the emitter. The holes immediately diffuse toward the collector because of the negative potential at that terminal.

The need for the extremely close spacing between the emitter and collector is now apparent. Many of the holes may meet with and be cancelled by the free electrons in the N-type material. Therefore, the flow path between the emitter and collector must be small to keep the hole and electron recombinations to a minimum.

At the collector electrode, the potential hill produced by the surface-bound electrons limits the current flow. However, holes that reach

Fig. 2-3 (top). Magnified view of hole and electron flow into point-contact transistor.

Fig. 2-4 (bottom). N-type and P-type point-contact transistor connections.

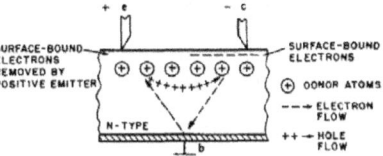

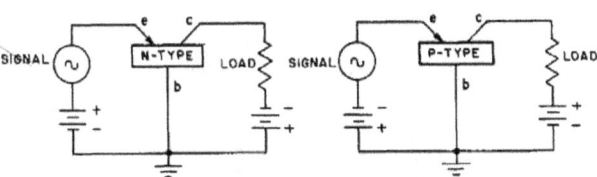

the collector area combine with the surface-bound electrons and reduce the potential hill. This permits the collector to inject more electrons into the germanium, thus increasing the collector current.

Holes travel through the transistor from emitter to collector in many indirect paths. The holes set up a net positive space charge in the areas of their flow paths, due to the combined effects of their positive charges. The resultant positive space charge attracts electrons from the more remote areas of the N-type transistor into the hole flow path between the collector and base, thus effectively increasing the electron flow. While some of the electrons emitted by the collector neutralize holes, the majority flow toward and enter the base terminal. The electrons which flow between the collector and the base are the collector current carriers.

Current, Resistance, Voltage, and Power Gains. In the average point-contact transistor, an increase in emitter current of one milliampere will cause an increase in collector current of 2.5 milliamperes. In physical terms, this indicates that one million holes injected by the emitter causes 2.5 million electrons to be injected by the collector. One million of the collector electrons neutralize the holes. The remaining million and a half electrons flow to the base.

The ratio of change in collector current to change in emitter current is called the current gain a (Alpha). Thus

$$a = \frac{i_c}{i_e}$$

where a = current amplification, i_e = change in emitter current, and i_c = resulting change in collector current.

In the typical case described above, $a = \dfrac{2.5 \text{ ma}}{1 \text{ ma}} = 2.5$.

At first glance, the current gain factor of a transistor is disappointingly low when compared with the amplification factor of a vacuum tube. However, another consideration enters the picture: The input resistance between the emitter and base is relatively low (300 ohms is a typical value), while the output resistance between collector and base is relatively high (20,000 ohms is typical). Thus, in addition to the current gain, the transistor has another gain characteristic, namely the ratio of output resistance to input resistance. For the typical point-contact transistor, the resistance gain is $\dfrac{20,000}{300} = 67$.

Since the input voltage is the product of the emitter current and the input resistance, and the output voltage is the product of the collector current and the output resistance, the transistor voltage gain equals the current gain times the resistance gain.

$$\text{Voltage gain} = \frac{e_o}{e_i} = \frac{i_c r_o}{i_e r_i} = a \, \frac{r_o}{r_i}$$

where: e_i = input voltage, e_o = output voltage,
  i_e = emitter current, i_c = collector current,
  a = current gain, r_o = output resistance, and
  r_i = input resistance.

For the typical case under consideration, the voltage gain equals $2.5 \times 67 = 167.5$. Furthermore, since the input power is the product of the input voltage and the emitter current, and the output power is the product of the output voltage and collector current, the *transistor power gain equals the current gain squared times the resistance gain.*

$$\text{Power gain} = \frac{e_o i_c}{e_i i_e} = a \, \frac{r_o}{r_i} \left(\frac{i_c}{i_e} \right) = a^2 \, \frac{r_o}{r_i}$$

For the typical transistor, the power gain equals $(2.5)^2 (67) = 419$.

P-Type Transistor. The P-type point-contact transistor operates similarly to the N-type unit, except that the emitter and collector battery polarities are reversed. Fig. 2-4 illustrates the essential difference between the battery connections for the two types of point-contact transistors.

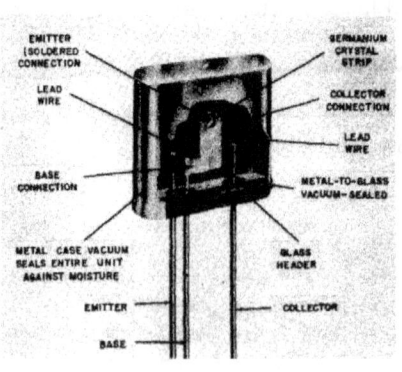

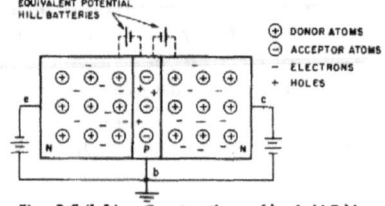

Fig. 2-5 (left). Construction of basic N-P-N junction transistor. Courtesy CBS-Hytron.

Fig. 2-6 (right). Basic N-P-N transistor.

Junction Transistors

Construction and Operation. In Chapter 1, it was observed that a combination of P- and N-type germanium form a P-N junction. In effect, this combination produces a germanium diode. The germanium diode has been incorporated in television circuits for several years to serve as second detectors, and has been used in other circuits where its excellent rectifying characteristic is useful.

Consider now the effect of combining two germanium diodes into one unit and, for further simplification, make the P-type section common to both. This new device, illustrated by Fig. 2-5, is the basic *N-P-N junction transistor.* In actual construction, the P section is very narrow as compared to the strips of N-type germanium. As will be seen later, the narrow middle section is required for proper transistor operation.

While the point-contact transistor required the relatively high-resistance point contacts used for the emitter and collector electrodes, all junction transistor electrodes are soldered to their respective sections, and make low resistance contact. The designations of the electrodes are the same as for the point-contact transistor, namely: the emitter (biased for forward or high conductivity), the collector (biased for reverse or low conductivity), and the base (which connects to the common P-junction area).

Although the junction transistor is a physical combination of two germanium diodes, conduction in the transistor is decidedly different from that in the diode or point-contact transistor. Observe that in Fig. 2-6 the negative potential at the emitter electrode pushes the free electrons towards the P-N junction. At the junction, as discussed in Chapter 1, a potential hill is set up by the action of the fixed donor and acceptor atoms. Since the emitter battery acts to flatten this emitter-base potential hill, a number of electrons pass this barrier and enter the P base region. The number of electrons crossing the barrier is proportional to the value of emitter battery potential. Some of these electrons combine with holes in the P base region, but most pass through and enter the N collector region. The loss of electrons in the P base region remains low (approximately five percent) because: (1) the base section is thin, and (2) the potential hill at the collector-base junction acts to accelerate the electrons into the N collector region. In the N region, the electrons are attracted to the positive collector.

P-N-P Transistor Operation. A *P-N-P junction transistor,* shown in Fig. 2-7, is formed by sandwiching a thin layer between two relatively thick P areas. As in the case of the N-P-N junction transistor, the electrode on the left is designated the emitter, the electrode on the right is designated the collector, and the common electrode is designated the base. However, the polarities of the potential hills formed are opposite those formed in the N-P-N junction transistor. In order to adhere to the general rules of biasing the emitter in the low resistance direction, and biasing the collector in the high resistance direction, the polarities of the external bias batteries are also reversed.

Conduction in the P-N-P junction transistor is similar to that in the N-P-N type. The holes in the P emitter region are repelled by the positive battery electrode toward the P-N junction. Since this potential hill is reduced by the emitter bias, a number of holes enters the base N area. A small number of holes (approximately five percent) is lost by combination with electrons within this area, and the rest move toward the collector, aided by the action of the collector-base potential hill. As each hole reaches the collector electrode, the collector emits an electron to neutralize the hole. For each hole that is lost by combination within the base or collector areas, an electron from one of the covalent bonds near the emitter electrode enters that terminal, thus forming a new hole in the vicinity of the emitter. The new holes immediately move toward the junction area. Thus a continuous flow of holes from the emitter to collector is maintained. It is evident that in both types of junction transistors discussed, the collector current is less than the emitter current by a factor proportional to the number of hole-electron recombinations that take place in the base junction area.

This analysis of the junction transistor leads to the following general observations:

1. The major current carriers in the N-P-N junction transistors are electrons.

2. The major current carriers in the P-N-P junction transistors are holes.

3. The collector current in either type of junction transistor is less than the emitter current because of the recombinations of holes and electrons in the base junction area. As an example, a typical rate of recombination is five percent. If an emitter current of one milliampere is assumed, the collector current is 1 ma. − (1 ma. x .05) = 0.95 ma.

Gain Factors. Since the current gain $\left(\alpha = \dfrac{i_c}{i_e}\right)$ of the junction transistor is always less than one, it might be expected that its voltage gain will be less than that of the typical point-contact transistor. In an actual case, however, the voltage gain of the junction transistor is considerably

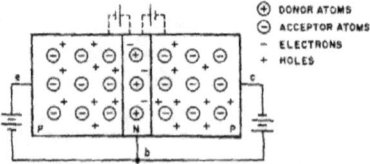

DONOR ATOMS
ACCEPTOR ATOMS
ELECTRONS
HOLES

Fig. 2-7. Basic P-N-P transistor.

larger than that of the point-contact type. Since the voltage gain is the product of the current gain and the resistance gain, it must be expected that the resistance gain is large. Typical values of emitter input and collector output resistances are 500 ohms and 1 megohm, respectively. Thus, the voltage gain

$$VG = a\frac{r_o}{r_i} = .95\left(\frac{1,000,000}{500}\right) = 1,900$$

$$PG = a^2\left(\frac{r_o}{r_i}\right) = (.95)^2\left(\frac{1,000,000}{500}\right) = 1,805$$

The high voltage and power gains of the junction transistor as compared to the point-contact transistor are due primarily to the high collector resistance.

Transistor Comparisons. To understand the factors which cause the relatively high collector resistance of the junction transistor as compared to the relatively low collector resistance of the point-contact transistor requires the aid of typical collector current-voltage characteristics. Figure 2-8 (A) illustrates the V_c-I_c characteristic of a typical point-contact transistor. As the collector voltage is raised above 5 volts, the current continues to increase, although at a diminishing rate due to the lack of available electrons in the transistor. As discussed previously, the holes in the point-contact transistor set up a positive space charge in the vicinity of their flow paths, attracting electrons from the more remote areas of the pellet. Thus, the electrons available for collector current decrease gradually.

Figure 2-8 (B) illustrates the V_c-I_c characteristic of a typical junction transistor. Here the V_c-I_c characteristic again follows an Ohm's law relationship at small values of collector voltage. The point of electron exhaustion is reached very abruptly, since there is no hole space-charge effect in the junction transistor to increase the available supply of electrons. After the critical voltage is attained, a large increase in collector

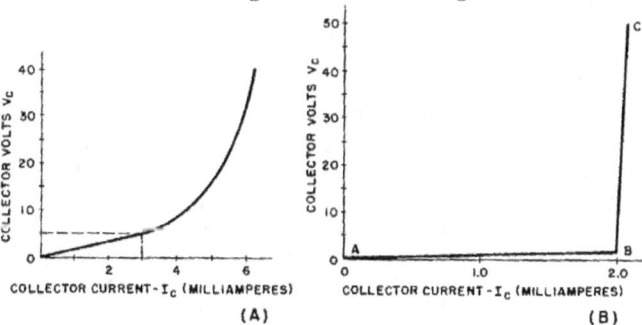

Fig. 2-8.   (A) Typical point-contact V_c-I_c characteristic.   (B) Typical junction V_c-I_c characteristic.

voltage causes only a very small increase in collector current. The collector resistance is equal to the change in collector voltage divided by the resulting change in collector current: $r_c = \dfrac{\Delta e_c}{\Delta i_c}$. For the typical junction transistor characteristic illustrated, the collector resistance from point A to B is $\dfrac{0.5}{2 \times 10^{-3}} = 250$ ohms, and from point B to C the collector resistance is $\dfrac{50}{.05 \times 10^{-3}} = 1,000,000$ ohms.

Because of the large collector resistance, and the resultant high resistance gain, the junction transistor is capable of far greater voltage and power gains than the point-contact types. Commercially available transistors with collector resistances in the neighborhood of 3 megohms are common; silicon-type junction transistors are inherently capable of far greater values.

The basic transistors have an upper frequency limit due to the small but finite time it takes the current carriers to move from one electrode to another. This limit, called the "alpha cutoff frequency," defines the point at which the gain is 3 db down from its low frequency value. The frequency response characteristics of transistors are considered more fully in Chapter 7.

In this chapter and those that follow, the germanium point-contact and junction transistors are considered at great length. This is not intended to create an impression that the entire field of semiconductors is limited to these two fundamental types. However, since their characteristics are basic to other semiconductor devices, a thorough understanding of the prototypes is essential. At this time, it appears that an unlimited number of variations of the original transistors is possible. Several of the more significant devices will now be considered.

The P-N Junction Photocell

Figure 2-9 (A) illustrates the essential construction and connections for the P-N junction photocell. The photocell is connected in series with a battery and a load resistor. The cell is biased by the battery in the reverse direction. Under these conditions, and with no light striking the P-N junction, approximately ten microamperes of current flow. The current value is low at this time because of the high resistance of the junction. However, when light strikes the P-N junction, the load current increases at a rate proportional to the light intensity.

These characteristics are illustrated by the typical operating curves shown in Fig. 2-9 (B). Notice that increasing the voltage from 20 to 100 volts, while holding the light constant, increases the current by less than 10 microamperes. However, increasing the light intensity from 3 to 6 millilumens increases the current approximately 100 microamperes.

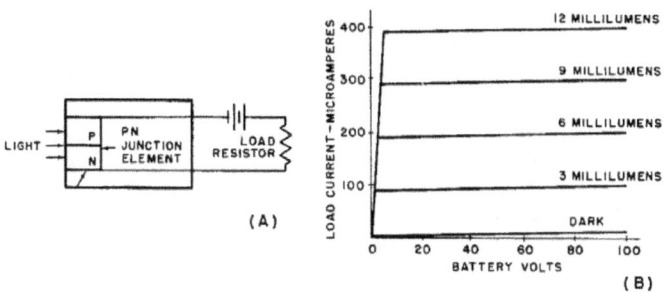

Fig. 2-9. (A) P-N junction photocell construction. (B) Typical junction photocell operating curves.

Basically, light from any source is composed of tiny particles of energy called *photons*. Thus, when light strikes the P-N junction element, in effect photons are bombarding the surface of the element and their energy is being absorbed by the germanium. The total energy absorbed is sufficient to disrupt some of the covalent bonds in the element, thereby creating free electrons and holes in the germanium, and increasing the number of available current carriers. When the light is removed, the current decreases rapidly because of the recombination of holes and electrons.

Wide-Spaced Transistors

It was noted previously that the emitter and collector contacts of the point-contact transistor must be closely spaced for normal transistor action, since the functioning of this transistor requires an intense electric field. In addition, the frequency response of this transistor decreases rapidly with increased contact spacing. Theoretically, the frequency operating band varies inversely as the cube of the contact spacing. In spite of this, it has been found that wide-spaced transistors have some novel and useful characteristics. When germanium having lower conductivity (fewer impurity atoms) is used, an increase in the normal contact spacing from .002 inch to as much as .015 inch has no effect on the transistor current and power gains. At the same time, the effect of the emitter voltage on the collector current is decreased, due to increased spacing.

The ratio of change in emitter voltage to the resulting change in collector current defines the backward transfer or feedback resistance. The feedback resistance in a transistor acts similarly to the positive feedback parameter in a vacuum tube circuit. (The feedback resistance and other related transistor characteristics are considered in greater detail in Chapter 3.) The feedback resistance of a transistor with a normal .002 inch contact spacing is about 200 ohms. This resistance is reduced to approximately 50 ohms when the contact spacing is increased to .015

inch. This low value insures circuit stability at relatively high values of power gain.

Germanium with higher than normal resistivity is used to compensate for the narrowing of the usable frequency limits by the wide contact spacing. Despite this, the usable frequency range is reduced to about 1/50 of its normal value. The increased contact spacing has little effect upon other transistor characteristics.

The P-N-P-N Transistor

Figure 2-10 illustrates the construction of the P-N-P-N junction transistor. This transistor, unlike the P-N-P or N-P-N junction transistor, is capable of a current gain. For satisfactory operation, both of the central P and N regions must be narrow.

In operation, the holes move in the direction from emitter to collector, but are trapped by the third potential hill in the collector area. The holes pile up at this barrier, and their cumulative positive space charge reduces the effect of the potential hill. As a result, electrons from the collector area encounter a decreased resistance at the junction and are able to flow into the central P region. Some electrons are lost through combinations with holes, but most of them, aided by the action of potential hill number 2, move into the middle N region and enter the base.

The P-N-P-N construction, because of the space charge effect of the holes, allows the current gain to reach values in the vicinity of 20. In comparison it must be remembered that the current gain of the prototype junction transistor is inherently limited to values less than one.

Transistor Tetrode

The frequency response of the conventional junction transistor is limited by several factors. First, the frequency cutoff (the frequency at which the current gain drops sharply) is inversely proportional to both the base resistance and to the square of the thickness of the junction layer. In addition, the frequency cutoff is also inversely proportional to the collector junction capacitance, considered only at high frequencies. Figures 2-11 (A) and 2-11 (B) illustrate the structural and symbolic representations of the junction transistor tetrode.

In this transistor, a fourth electrode, designated as b_2, is included. The fourth electrode is connected to the P junction layer in the same manner as the conventional base electrode, but the connection is made on the opposite side of the layer. The base resistance is reduced sub-

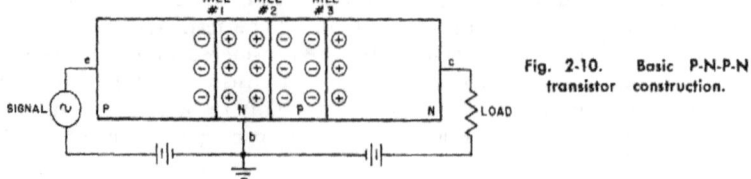

Fig. 2-10.   Basic P-N-P-N transistor construction.

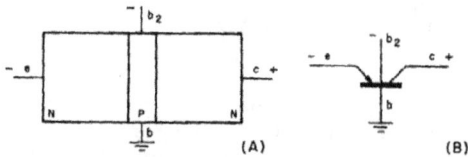

Fig. 2-11. Tetrode junction
transistor: (A) structural
representation, (B) symbolic
representation.

stantially when a negative bias is applied to the second base electrode. The bias prevents that part of the emitter junction which is near b_2 from emitting electrons into the P layer. Thus all of the transistor action takes place near the base. This effectively reduces the base resistance; as a result, the frequency response increases.

For proper operation, the second base electrode is biased to about —6 volts with respect to the base. The resulting bias current is approximately one milliampere. In a typical case this bias reduces the base resistance from 1,000 to 40 ohms; the change in emitter resistance is negligible. The current gain is reduced from .95 to .75, and the collector resistance is reduced from 3.0 to 1.5 megohms. The frequency response cutoff is increased from 0.5 to 5 megacycles. Thus, an increased bandwidth is obtained at the expense of lower available gain.

The effect of the junction area thickness is decreased by using very thin P layers (roughly .0005 inch). The collector junction capacitance is reduced by decreasing the collector junction area.

Chapter 3
THE GROUNDED BASE TRANSISTOR

This chapter deals with basic four-terminal analysis in general, and the specific application of four-terminal network analysis to the transistor. Hence, the important characteristics of the transistor, including the open-circuit parameters, the current gain, the voltage gain, the power gain, and the conditions for image input and output resistance match are derived. The basic principles and connections for measuring transistor characteristics are discussed, and a comparison between the transistor and the electron tube is considered.

While the mathematics involved in the analysis of the transistor has been held to a bare minimum, some readers may be dismayed at what appears to be an excessive number of derivations. It cannot be overemphasized, however, that a thorough understanding of the transistor requires a general knowledge of the mathematical analysis leading to the major design formulas. These important design equations are noted by an asterisk (*).

Four-Terminal Networks

In all types of engineering circuit design, it is frequently convenient to represent a device by an electrical equivalent. This invariably eases the task of optimizing the design, since the device is, in effect, reduced to a simpler equivalent form. One of the most useful methods of equivalent representations is by means of the *four-terminal network*.

The four-terminal network (also called a coupling network, or two-terminal pair network) is shown in Fig. 3-1. Terminals *a* and *b* represent the input to the network and terminals *c* and *d* the output. The network itself, which represents the equivalent of a device or any combination of devices, is located between the input and output terminals, and is considered sealed, so that electrical measurements can be made only at the input and output terminals.

The sealed network may be, and often is, very complex. As an example, consider the case of relating the acoustical input to a microphone in a multi-link transmission circuit to the acoustical output of a receiver. This system involves transmission lines, electronic circuts, acoustical, electrical, and mechanical power and transducers. In the four-terminal method of analysis, however, the complete intermediate system between the microphone input and the receiver output is represented by the sealed box.

The advantage of this type of representation is that only one basic analysis of a particular device or system is required. Once accomplished, problems involving the same system or device are a matter of routine and become simple substitutions of numbers. For electronic devices, other

Fig. 3-1. Four-terminal network, conventional designation.

advantages are that the basic equivalent circuit can be modified to include the effects of high-frequency operation, and that the equivalent circuit invariably contains a minimum number of parameters which can be directly related to external measurements.

Four-terminal networks are divided into two general classifications: active and passive. Passive networks are those that contain no source of energy within the sealed box; currents and voltages within the box are a result of the application of energy to the external terminals. Examples of passive networks include filters, attenuators, and transmission lines. Active networks, on the other hand, do contain internal sources of energy. Examples of these, therefore, include all types of amplifying devices, including the transistor. Although the conventional transistor has but three external connections, four-terminal network analysis is applicable because one of the electrodes is common to both the input and output circuits.

The performance of the transistor can be completely defined by the voltage and current measured at the input and output terminals. Actually only two of the four values are independent, because if any two are specified, the other two values are automatically determined. This situation is exactly the same as that in the conventional triode electron tube, where the four values are the grid current, grid voltage, plate current, and plate voltage. The grid and plate voltages of a tube are usually considered the independent variables, and their respective currents then become the dependent variables.

General Four-Terminal Network Analysis

The general four-terminal active network is fully described by the relationship between the input and output currents and voltages. Referring to Fig. 3-1, the general voltage (loop) equations are:

$$E_1 = Z_{11}I_1 + Z_{12}I_2$$

and

$$E_2 = Z_{21}I_1 + Z_{22}I_2$$

where Z_{11} is the input impedance with the output open.

$Z_{11} = E_1/I_1$, when $I_2 = 0$.

Z_{12} is the feedback or reverse transfer impedance with the input open.

$Z_{12} = E_1/I_2$, when $I_1 = 0$.

Z_{21} is the forward transfer impedance with the output open.

$Z_{21} = E_2/I_1$, when $I_2 = 0$.

Z_{22} is the output impedance with the input open.

$Z_{22} = E_2/I_2$, when $I_1 = 0$.

The equivalent current (nodal) equations are
$$I_1 = Y_{11}E_1 + Y_{12}E_2$$
and
$$I_2 = Y_{21}E_1 + Y_{22}E_2$$
where Y_{11} is the input admittance with the output shorted.

$Y_{11} = I_1/E_1$, where $E_2 = 0$.

Y_{12} is the feedback or reverse transfer admittance with the input shorted.

$Y_{12} = I_1/E_2$, when $E_1 = O$.

Y_{21} is the forward transfer admittance with the output shorted.

$Y_{21} = I_2/E_1$, when $E_2 = O$.

Y_{22} is the output admittance with the input shorted.

$Y_{22} = I_2/E_2$, when $E_1 = 0$.

Amplification factors are the best general index of an active network. Since the general case may have amplification in both directions, definitions are included for forward and reverse directions.

The forward current amplification factor, a_{21}, is equal to the negative ratio of the current at the shorted output terminals to the current at the input terminals.

$$a_{21} = -\left(\frac{I_2}{I_1}\right) \text{when } E_2 = 0$$

Then $0 = E_2 = Z_{21}I_1 + Z_{22}I_2$.

Solving these equations $a_{21} = -\left(\frac{I_2}{I_1}\right) = -\frac{Z_{21}}{Z_{22}}$ and in terms of admittance

$$a_{21} = -\left(\frac{Y_{21}}{Y_{11}}\right)$$

The reverse current amplification factor, a_{12}, is equal to the negative ratio of the current at the shorted input terminals to the current at the output terrminals:

$$a_{12} = -\left(\frac{I_1}{I_2}\right), \text{ when } E_1 = 0$$

Then $0 = E_1 = Z_{11}I_1 + Z_{12}I_2$.

Solving as before, $a_{12} = -\left(\frac{I_1}{I_2}\right) = -\frac{Z_{12}}{Z_{11}}$, and in terms of admittances

$$a_{12} = -\left(\frac{Y_{12}}{Y_{22}}\right)$$

The forward voltage amplification factor, μ_{21}, is equal to the ratio of the open circuit output voltage to the input voltage. $\mu_{21} = \frac{E_2}{E_1}$, when $I_2 = 0$. On this basis, $E_1 = Z_{11}I_1$ and $E_2 = Z_{21}I_1$.

Thus $\mu_{21} = \dfrac{E_2}{E_1} = \dfrac{Z_{21}}{Z_{11}}$ and on an admittance basis $\mu_{21} = -\left(\dfrac{Y_{21}}{Y_{22}}\right)$.

The reverse voltage amplification factor μ_{12} is equal to the ratio of the open circuit input voltage to the output voltage. $\mu_{12} = \dfrac{E_1}{E_2}$ when $I_1 = 0$. Then $E_1 = Z_{12}I_2$ and $E_2 = Z_{22}I_2$. Thus $\mu_{12} = \dfrac{E_1}{E_2} = \dfrac{Z_{12}}{Z_{22}}$

In terms of admittance $\mu_{12} = -\left(\dfrac{Y_{12}}{Y_{11}}\right)$

Vacuum-Tube Analysis on a Four-Terminal Basis

Figure 3-2 (A) illustrates the familiar case of a conventional grounded-cathode triode operated at low frequencies with its control grid biased sufficiently negative so that no grid current flows. (It should be noted at this point that the current arrows in this diagram and those that follow indicate the direction of electron flow.) The applied grid signal causes a voltage μe_g to appear in series with the plate resistance r_p. Since the grid current i_g is zero, the network is completely described by a single equation:

$$\mu e_g = i_p r_p + e_p$$

The four-pole equivalent network for this same circuit when the grid draws current is illustrated by Fig. 2 (B). In this case, the grid voltage acts across a series circuit consisting of the voltage $\mu_p e_p$ and the grid resistance r_g. The term μ_p equals the reverse voltage amplification factor: $\mu_p = \dfrac{e_g}{e_p}$. As in the previous case, the grid signal voltage causes a voltage μe_g to appear in series with the plate resistance. Since there are two voltage loops in the case when the grid is driven positive, two equations are required to describe the network completely; these are

$$\mu e_g = i_p r_p + e_p$$
$$e_g = i_g r_g + \mu_p e_p$$

This analysis of triode vacuum tubes on a four-pole basis is not limited to the grounded-cathode operation of these tubes. The choice of this type of operation is dictated on the basis of reader familiarity with

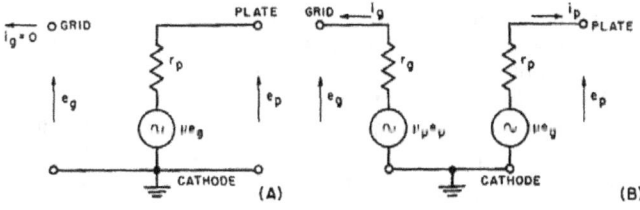

Fig. 3-2. Equivalent circuit of a triode: (A) with negative grid bias, (B) with positive grid bias.

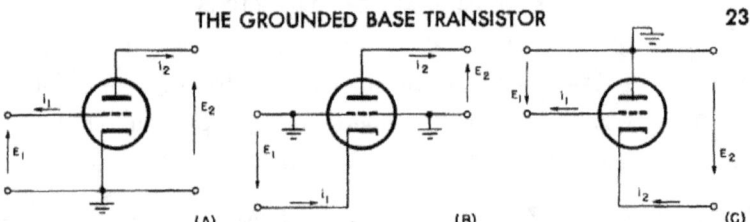

Fig. 3-3. Four-terminal network ground connection: (A) cathode, (B) grid, (C) plate.

the circuit. The grounded-grid and grounded-plate connections (which have useful counterparts in transistor circuitry) may be analyzed in similar fashion. The basic four-terminal current-voltage relationships for all three cases are illustrated in Fig. 3-3.

Vacuum Tubes Compared with Transistors

Representation of a vacuum-tube circuit by an equivalent circuit which includes its transconductance, amplification factors, plate resistance, and grid resistance is particularly useful in design applications. This treatment greatly simplifies analysis in those applications of the tube's operating characteristics where a linear approximation is valid. A similar type of linear analysis is applicable to the operation of the transistor. As will be seen shortly, transistor parameters correspond closely to tube parameters. The main factor contributing to differences between tube and transistor characteristics is that the transistor is primarily a current operated device, while a vacuum tube is a voltage operated device.

When the grid of a vacuum tube is held negative with respect to the cathode, only three tube variables exist, since the grid current is zero. The transistor, however, always has four variables. As a result, four independent parameters are necessary to specify its characteristics completely. The analysis that follows is based on small signal inputs which satisfy the requirements of linearity. For this reason, the resulting parameters are called the *small-signal parameters.*

In the transistor, both the input and output currents and voltages are significant. In addition, it is possible to have two or more sets of currents for one set of voltages. This situation is somewhat similar to that existing in a vacuum tube that draws grid current, in which there may be two possible grid voltages for a given set of grid and plate currents. In the transistor, there can only be one set of voltages for a specified pair of input and output currents. This reason governs the choice of current as the independent variable in transistor work as opposed to the choise of voltage in the representation of vacuum-tube characteristics.

In vacuum tubes the input grid voltage is plotted against a plate characteristic, because the output voltage is approximately a linear function of the grid voltage. In transistor circuitry, a similar curve can be formed by plotting the collector voltage as a function of collector current for a fixed value of input current. Note again that the transistor input current is selected as the independent variable rather than input voltage. The grounded-cathode vacuum tube is a voltage amplifying device having a high input impedance and a relatively low output impedance. Its equivalent transistor circuit, the grounded emitter transistor, is a current amplifying device with a low input impedance and a relatively high output impedance.

Several types of equivalent circuits can be used to represent the transistor under small signal conditions. Figure 3-4 represents only three of the many possibilities. The indicated circuits are equivalent in that they all give the same performance for any given set of input and output characteristics. Examples (B) and (C) are particularly well suited to transistor application because the resulting parameters are of significance in transistor physics. In addition, the parameters are readily measured, are usually positive, and are not extremely dependent on the exact operating point chosen. The significance of the impedance parameters is covered later in the chapter.

The derivations of these equivalent circuits are based on the relationship between the input and output currents and voltages. For example, assume that for the sealed network of Fig. 3-1 the input and output resistances remain constant with frequency and are each equal to 200 ohms. Then the network may be a shunt resistor equal to 200 ohms (Fig. 3-5A), a "T" pad of three equal 100-ohm resistors (Fig. 3-5B), a "pi" pad of three equal 300-ohm resistors (Fig. 3-5C), or any other combination meet-

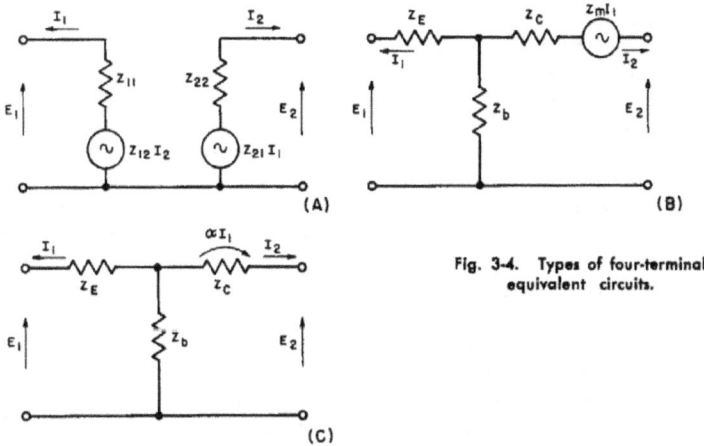

Fig. 3-4. Types of four-terminal equivalent circuits.

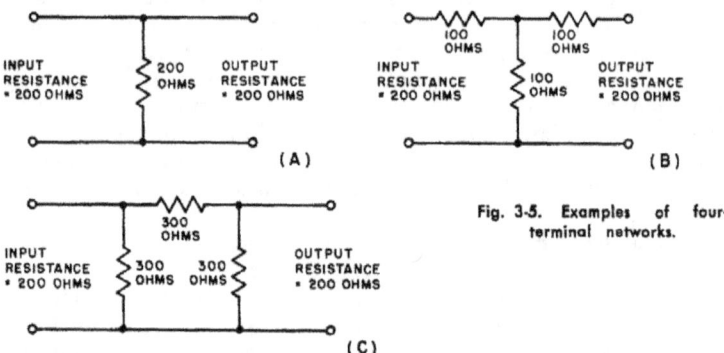

Fig. 3-5. Examples of four-terminal networks.

ing the required input and output characteristics. The derivations of active networks are admittedly more complicated than this simple example, but the basic principles are exactly the same.

Four-Terminal Analysis of Transistors

Like the vacuum tube triode, the transistor has useful properties in any of the three possible connections: grounded base, grounded emitter, and grounded collector. Most of the present literature starts with the grounded base connection because this configuration is the most convenient for describing transistor physics. In circuit work, however, the grounded emitter connection is most popular because it provides maximum obtainable power gain for a specified transistor and is well suited to cascading without impedance-matching devices. The vacuum tube counterpart of this circuit, the grounded cathode connection, also produces maximum power gain and is adaptable to cascading without impedance-matching devices.

Typical characteristics for a junction transistor in grounded base connection are shown in Fig. 3-6 (A). Since the collector current is the independent variable, it is plotted along the abscissa, in apposition to the method used in plotting vacuum tube characteristics. Notice the similarity between the junction transistor characteristics in Fig. 3-6 (A) and those of the typical triode vacuum tube illustrated in Fig. 3-6 (B). Based on this similarity, it is reasonable to assume that the transistor collector voltage, collector current, and emitter current can be compared with the plate current, plate voltage, and grid voltage of a triode vacuum tube.

Examining the tube characteristics, it is seen that a signal applied to the grid shifts the plate voltage along the load line. The numerical plate voltage shift caused by a change of one volt in the grid voltage is defined as the amplification factor μ of the tube. In a like manner, applying a signal to the transistor emitter shifts the collector current along the load line. The numerical shift in collector current caused by a

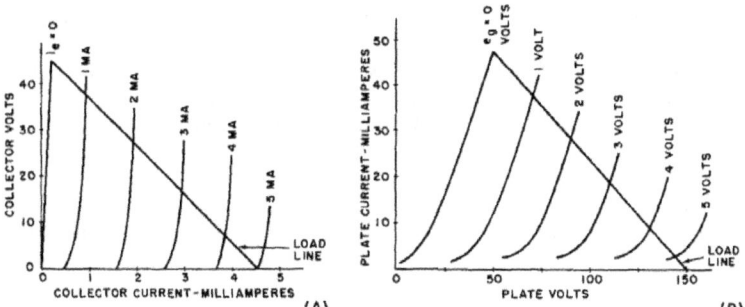

Fig. 3-6. (A) Typical junction-transistor characteristics. (B) Typical vacuum-tube characteristics.

change in emitter current of one milliampere is defined as the current amplification factor a of the transistor. The current amplificaton factor of a transistor, then, corresponds to the voltage amplification factor of a vacuum tube.

Insofar as input characteristics are concerned, the vacuum tube normally operates with its grid biased in the reverse or high resistance direction, while the transistor operates with the emitter biased in the forward or low resistance direction. In the output circuits, a similar relationship exists. The plate of a vacuum tube is biased in the forward direction, while the transistor collector is biased in the reverse direction. These biasing conditions produce the high input and low output impedances in the vacuum tube circuit, and the low input and high output impedances in the transistor circuit. This re-emphasizes the basic difference between the vacuum tube and the transistor: the vacuum tube is a voltage controlled device, while the transistor is a current controlled device.

Equivalent Passive "T" Network. In the analysis of the transistor on a four-terminal basis, the entire device is treated as a sealed box with three external terminals $e, b, c,$ designating the emitter, base, and collector, respectively. This basic four-terminal network is illustrated in Fig. 3-7 (A), in which the input signal is applied between emitter and base. The input signal E_g is taken from a signal generator that has an internal resistance R_g. The output circuit is between the collector and the common base and consists of a load resistance R_L. In the small-signal analysis which follows, it is assumed that the transistor is biased in the linear region of its operating characteristics. It is also assumed that the operating frequency is low enough so that the transistor parameters may be considered pure resistances, and the capacitive junction effects may be considered negligible.

The simplest method of approaching the analysis of the equivalent transistor circuit is by an equivalent "T" network with no internal generating sources (passive basis). This circuit is illustrated in Fig. 3-7 (B). Under these conditions, the transistor parameters can be completely specified by the following terminal measurements:

A. Input resistance with output terminals open,

$$r_{11} = \frac{e_i}{i_e}$$

when $i_c = 0$, $r_{11} = r_e + r_b$.

B. The forward transfer resistance with the output terminals open

$$r_{21} = \frac{e_o}{i_e}$$

when $i_c = 0$, $r_{21} = r_b$.

C. The output resistance with the input terminals open,

$$r_{22} = \frac{e_o}{i_c}$$

when $i_e = 0$, $r_{22} = r_c + r_b$.

D. The backward transfer resistance with the input terminals open

$$r_{12} = \frac{e_i}{i_c}$$

when $i_e = 0$, $r_{12} = r_b$.

Notice that the forward transfer resistance is equal to the backward transfer resistance. This is typical of a four-terminal passive network. In the practical case, then, it is only necessary to measure r_{12} or r_{21}.

Equivalent Active "T" Network. While the passive network serves as an interesting introduction to transistor analysis, it does not describe this device completely, because the transistor is known to be an active

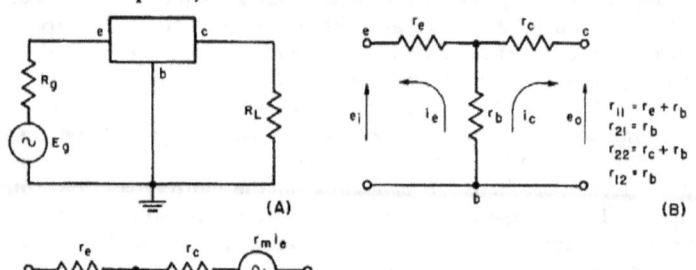

(A)   (B)

$r_{11} = r_e + r_b$
$r_{21} = r_b$
$r_{22} = r_c + r_b$
$r_{12} = r_b$

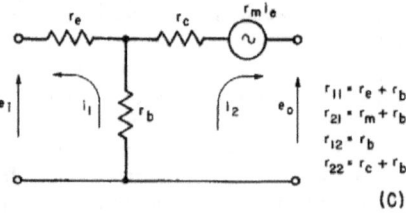

(C)

$r_{11} = r_e + r_b$
$r_{21} = r_m + r_b$
$r_{12} = r_b$
$r_{22} = r_c + r_b$

Fig. 3-7. (A) The basic circuit for transistor four-terminal network analysis. (B) Transistor equivalent "T" on a passive basis. (C) Transistor equivalent "T" on an active basis.

network. The equivalent circuit of the transistor can be represented in a number of ways; the most widely used configuration is illustrated in Fig. 3-7 (C). The basic difference between the equivalent circuits representing the active network and the passive network is the voltage source $e = r_m i_e$ inserted in the collector arm. In general, the passive network determines three of the characteristics of the active network. In the case under consideration, the input resistance r_{11}, the output resistance r_{22}, and the backward transfer resistance r_{12} are the same for the passive and active networks shown in Figs. 3-7 (B) and 3-7 (C). The only difference is the value of the forward transfer resistance r_{21}, which in the case of the passive network equals r_b, and in the active network equals $r_m + r_b$.

Thus, the equations for an active network under ideal conditions (that is, when the resistance of the signal source is zero, and the resistance of the load is infinite) become:

$$r_{11} = r_e + r_b$$
$$r_{12} = r_b$$
$$r_{21} = r_b + r_m$$
$$r_{22} = r_c + r_b$$

The four parameters in this active network are the emitter resistance r_e, the base resistance r_b, the collector resistance r_c, and the voltage source $r_m i_e$. The parameter r_m is represented as a resistance since it acts as the proportionality constant between the input emitter current and the resulting voltage source in the collector arm. The mathematical logic of the resistance r_m is easily derived as follows: In preceding chapters, the current gain a was defined as the ratio of the resulting change in collector current to a change in emitter current, $a = \dfrac{i_c}{i_e}$. The equivalent voltage introduced into the collector circuit is $e = i_c r_c$. Since $i_c = a i_e$, $e = a i_e r_c$. Since a is a dimensionless parameter, it can be related to the collector resistance by a resistance parameter r_m. Thus, $a = \dfrac{r_m}{r_c}$. Substituting this latter equality, $e = a i_e r_c = \dfrac{r_m}{r_c} i_e r_c = r_m i_e$

There is no phase inversion in the grounded base connection; a positive signal applied to the emitter produces an amplified positive signal of the same phase at the collector.

Measuring Circuits. Figure 3-8 illustrates the four basic circuits for measuring four-terminal parameters. The double subscript designations on the general resistance parameters of the four-terminal network (those designated r_{11}, r_{12}, r_{21}, and r_{22}) refer to the input terminal 1 and the output terminal 2. In addition, the first subscript refers to the voltage, and the second subscript refers to the current. For example, r_{12} is the ratio of the input voltage to the output current, while r_{21} is the ratio of

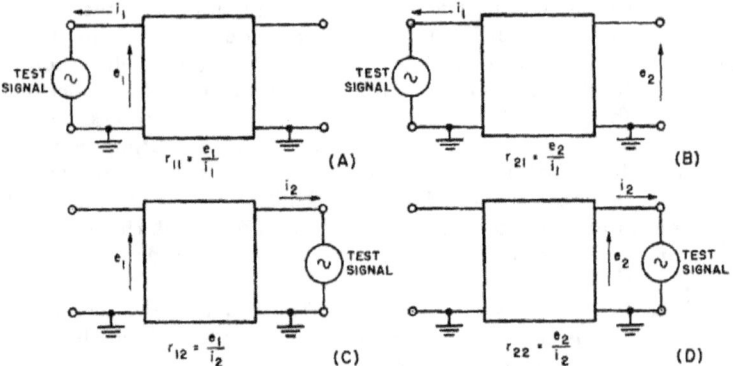

Fig. 3-8. Basic circuits for measuring four-terminal parameters.

the output voltage to the input current. These designations also indicate whether a test signal is applied to the input or output terminals, since the current will always be measured at the terminals where the test signal is applied.

There are several sources of error inherent in this use of small-signal inputs to evaluate the parameters of a transistor. The first error is due to the non-linearity of the characteristic curves. The larger the signal input, the greater the error. Thus, to make this error negligible, the signal must be held sufficiently small. In the practical case, the minimum useful signal is limited by the transistor thermal noise.

A second error results from the internal resistance of the signal source. In measuring any of the parameters, the amplitude of the input signal is assumed to be independent of the transistor resistance. However, this is true only if the source impedance is very much greater or very much smaller than the transistor input resistance. If a high resistance source is used, the magnitude of the resulting error is proportional to the ratio of the transistor input resistance and the resistance of the signal source. If a low resistance source is used, the error is proportional to the ratio of the internal resistance of the signal source and the input resistance of the transistor.

The third source of error, when small signal inputs are assumed, results from the shunting action of the input voltmeter and the input d-c bias supply on the input signal. The first effect may be made negligible by using a very high resistance voltmeter. The magnitude of the error caused by the shunting action of the d-c bias supply is proportional to the ratio of the transistor input resistance and the d-c supply resistance.

Errors are also introduced by the capacities between emitter and base, and between collector and base. These capacitance effects are comparable to the plate to grid and cathode to grid capacitances in a vacuum tube. In the audio frequency range, these errors are generally neglected.

The Grounded Base Connection

Equivalent Operating Circuit. At this point, the transistor equivalent circuit must be considered using a practical circuit, such as illustrated in Fig. 3-9. The signal generator E_g, having an internal resistance R_g, is connected between the emitter and the base. A load resistance R_L is connected between the collector and the common base. The input current is designated i_1, and for the common base connection is equal to the emitter current i_e. The collector output current is designated i_2. A cursory look at Fig. 3-9 makes it fairly evident that the input resistance r_{11} as seen by the signal generator depends to some extent on the value of the load resistance R_L, and the output resistance r_{22} as seen by the load resistance is determined to some extent by the value of the generator's internal resistance R_g. On a basis of Kirchoff's Law, the loop equations for the circuit of Fig. 3-9 are:

Input loop 1: $\quad E_g = i_1 (R_g + r_e + r_b) + i_2 r_b \qquad Eq. \ (3\text{-}1)$

Output loop 2: $\quad -r_m i_e = i_1 r_b + i_2 (r_c + r_b + R_L)$

Since $i_e = i_1$, then

$$O = i_1 (r_b + r_m) + i_2 (r_c + r_b + R_L) \qquad Eq. \ (3\text{-}2)$$

Since these two loop equations are independent, they may be solved simultaneously for the two unknown currents i_1 and i_2. Then

$$i_1 = \frac{E_g (r_b + r_c + R_L)}{(R_g + r_e + r_b)(r_b + r_c + R_L) - r_b (r_b + r_m)} \qquad Eq. \ (3\text{-}3)$$

$$i_2 = \frac{E_g (r_b + r_m)}{(R_g + r_e + r_b)(r_b + r_c + R_L) - r_b (r_b + r_m)} \qquad Eq. \ (3\text{-}4)$$

Under ideal conditions, namely, when R_g equals zero, and R_L is infinite, it was previously found that $r_{11} = r_e + r_b$, $r_{12} = r_b$, $r_{21} = r_b + r_m$, and $r_{22} = r_e + r_b$.

If these values are substituted in equations 3-3 and 3-4, i_1 and i_2 can be evaluated in terms of the ideal or open-circuit parameters.

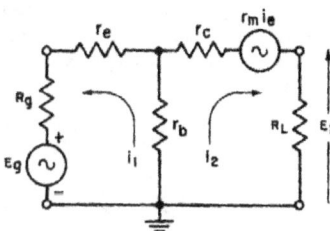
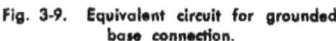
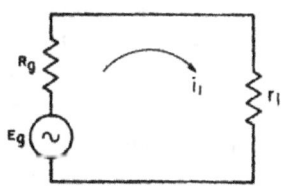

Fig. 3-9. Equivalent circuit for grounded base connection.

Fig. 3-10. Simplified transistor equivalent circuit for analysis of input resistance r_i.

$$i_1 = \frac{E_g (R_L + r_{22})}{(R_g + r_{11})(R_L + r_{22}) - r_{12}r_{21}} \qquad Eq. \ (3\text{-}5)$$

$$i_2 = \frac{E_g r_{21}}{(R_g + r_{11})(R_L + r_{22}) - r_{12}r_{21}} \qquad Eq. \ (3\text{-}6)$$

Current Gain. The current gain, $a = \dfrac{i_2}{i_1}$, when the circuit is working into the load R_L, becomes the ratio of equation *3-4* to equation *3-3*

$$a = \frac{r_b + r_m}{r_b + r_e + R_L} \qquad Eq. \ (3\text{-}7) \ *$$

and in terms of the open-circuit parameters, the current gain is the ratio of equation *3-6* to equation *3-5*

$$a = \frac{r_{21}}{R_L + r_{22}} \qquad Eq. \ (3\text{-}8) \ *$$

The current gain as derived in equations *3-7* and *3-8* indicates the effect of the load resistance R_L on a, but does not take into account the effect of mismatch between the signal generator resistance and the input resistance of the transistor. It is evident that the maximum current gain is obtained when the load resistance $R_L = O$. Thus, the maximum

$$a_0 = a = \frac{r_{21}}{r_{22}} = \frac{r_b + r_m}{r_b + r_e} \qquad Eq. \ (3\text{-}8A) \ *$$

Since r_b is very small in comparison with either r_m or r_e, it may be neglected in equation *3-8A*, and a rather accurate estimate of the maximum current gain is

$$a = a_0 = \frac{r_m}{r_e} \qquad Eq. \ (3\text{-}8B) \ *$$

A frequently used form for the current gain, which incorporates the maximum current gain a_0, is

$$a = \frac{a_0}{1 + \dfrac{R_L}{r_{22}}} \qquad Eq. \ (3\text{-}8C) \ *$$

Input Resistance, r_1. The input resistance of the grounded base transistor shown in Fig. 3-9 can now be computed in terms of the transistor parameters and the transistor four-terminal open-circuit parameters. Since the input resistance as seen by the signal generator is r_1, Fig 3-9 may be simplified as shown in Fig. 3-10. This series circuit is expressed

$$E_g = i_1 (R_g + r_1)$$

or

$$R_g + r_1 = \frac{E_g}{i_1} \qquad Eq. \ (3\text{-}9)$$

Substituting equation *3-3* for i_1

$$R_g + r_1 = \frac{E_g \left[(R_g + r_e + r_b)(r_b + r_c + R_L) - r_b(r_b + r_m) \right]}{E_g (r_b + r_c + R_L)} \quad Eq. \ (3\text{-}10)$$

$$= R_g + r_e + r_b - \left(\frac{r_b(r_b + r_m)}{r_b + r_c + R_L} \right) \qquad Eq. \ (3\text{-}11)$$

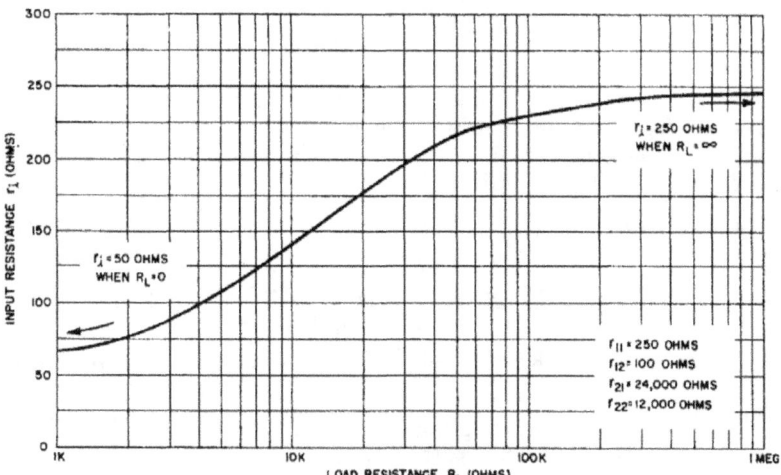

Fig. 3-11. Input resistance vs load resistance for typical point-contact transistor (grounded base).

then
$$r_i = r_e + r_b - \left(\frac{r_b (r_b + r_m)}{r_b + r_e + R_L} \right) \qquad Eq.\ (3\text{-}12)\ *$$

and in terms of the open-circuit parameters

$$r_i = r_{11} - \frac{r_{12} r_{21}}{r_{22} + R_L} \qquad Eq.\ (3\text{-}13)\ *$$

The effect of varying the load resistance on the input resistance can be best appreciated by examining Figs. 3-11 and 3-12, which illustrate the r_i *vs* R_L characteristics for typical point-contact, and junction transistors, respectively. For the typical point-contact transistor, $r_{11} = 250$ ohms, $r_{12} = 100$ ohms, $r_{21} = 24,000$ ohms, and $r_{22} = 12,000$ ohms. For the typical junction transistor, $r_{11} = 550$ ohms, $r_{12} = 500$ ohms, $r_{21} = 1,900,000$ ohms, and $r_{22} = 2,000,000$ ohms. Notice that in the case of the point-contact transistor, the transistor input resistance varies from 50 to 250 ohms as the load resistance changes from zero to infinity. The junction transistor input resistance varies from 75 to 550 ohms as the load resistance is varied from short-circuit to open-circuit conditions.

Output Resistance, r_o. The output resistance can be found in a similar manner. Consider Fig. 3-13 (A), which illustrates the equivalent circuit for analyzing the output resistance. The equations for the two loops on the basis of Kirchoff's law are:

Loop 1: $O = i_1 (R_g + r_e + r_b) + i_2 r_b$ *Eq. (3-14)*

Loop 2: $E_2 - r_m i_e = i_1 r_b + i_2 (r_e + r_b + R_L)$ *Eq. (3-15)*

Since $i_e = i_1$, then

$E_2 = i_1 (r_b + r_m) + i_2 (r_e + r_b + R_L)$ *Eq. (3-15A)*

Solving the two independent equations *3-14* and *3-15A* for the unknown load current,

$$i_2 = \frac{(R_g + r_e + r_b) E_2}{(R_g + r_e + r_b)(r_c + r_b + R_L) - r_b(r_m + r_b)} \qquad Eq. \ (3\text{-}16)$$

Looking back into the transistor, the generator E_2 with its internal resistance R_L sees the output resistance r_o. Again the circuit may be simplified as shown in Fig. 3-13 (B).

Then $E_2 = (R_L + r_o) i_2$ or $R_L + r_o = \dfrac{E_2}{i_2}$ $\qquad\qquad Eq. \ (3\text{-}17)$

Substituting equation *3-16* in *3-17*,

$$R_L + r_o = \frac{E_2 \left[(R_g + r_e + r_b)(r_c + r_b + R_L) - r_b(r_m + r_b) \right]}{E_2(r_b + r_c + R_g)} Eq. \ (3\text{-}18)$$

$$= r_c + r_b + R_L - \left(\frac{r_b(r_m + r_b)}{r_b + r_c + r_m} \right) \qquad\qquad Eq. \ (3\text{-}19)$$

Then $r_o = r_c + r_b - \left(\dfrac{r_b(r_m + r_b)}{r_b + r_c + r_m} \right)$ $\qquad\qquad Eq. \ (3\text{-}20) \ *$

In terms of the open-circuit parameters

$$r_o = r_{22} - \left(\frac{r_{12}r_{21}}{R_g + r_{11}} \right) \qquad\qquad Eq. \ (3\text{-}21) \ *$$

The latter equation indicates that the output resistance depends to some extent on the value of the signal generator input resistance. The variation of r_o *vs* R_g is illustrated in Figs. 3-14 and 3-15 (for the same point-contact and junction transistors considered in the preceding section). In the case of the typical point-contact transistor, the transistor

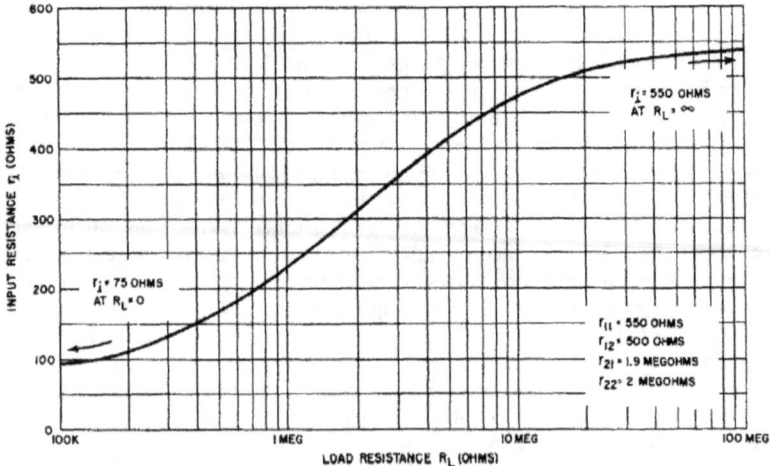

Fig. 3-12. Input resistance vs load resistance for typical junction transistor (grounded base).

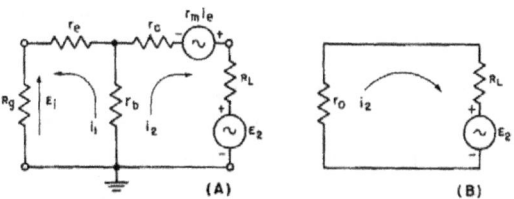

Fig. 3-13. Analysis of output resistance r_o: (A) equivalent circuit, (B) simplified circuit.

output resistance varies from 2,400 to 12,000 ohms as the signal generator internal resistance increases from zero to infinity. The junction transistor output resistance varies from 270,000 to 2,000,000 ohms as the signal generator internal resistance increases from zero to infinity.

Voltage Gain VG. Looking again at Fig. 3-9, it is seen that the voltage gain $VG = \dfrac{E_2}{E_g}$. Since $E_2 = i_2 R_L$ and $E_g = i_1(R_g + r_1)$,

$$VG = \left(\frac{i_2}{i_1}\right)\left(\frac{R_L}{R_g + r_1}\right) \qquad\qquad Eq.\ (3\text{-}22)$$

Since $\dfrac{i_2}{i_1} = a$, if equation 3-8 is substituted for $\dfrac{i_2}{i_1}$ in equation 3-22 and if equation 3-13 is substituted for r_1 in equation 3-22 the volt-

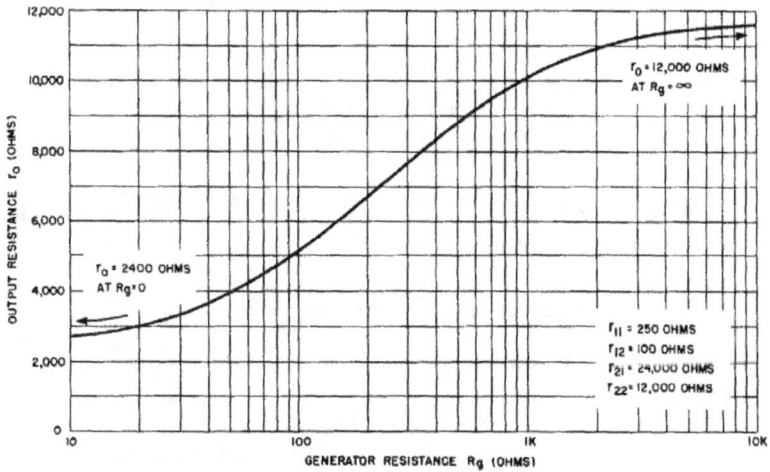

Fig. 3-14. Output resistance vs generator resistance for typical point-contact transistor (grounded base).

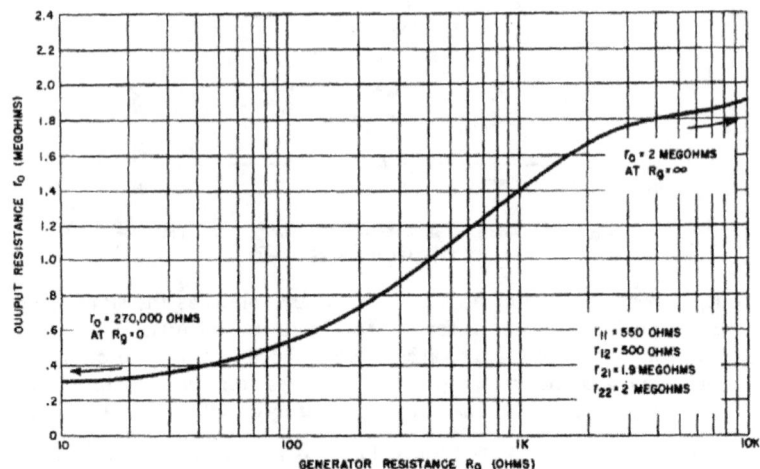

Fig. 3-15. Output resistance vs generator resistance for typical junction transistor (grounded base).

age gain becomes:

$$VG = \frac{r_{21}R_L}{(R_L + r_{22})\left[R_g + r_{11} - \left(\dfrac{r_{12}r_{21}}{r_{22} + R_L}\right)\right]} \qquad Eq.\ (3\text{-}23)$$

$$VG = \frac{r_{21}R_L}{(R_L + r_{22})(R_g + r_{11}) - r_{12}r_{21}} \qquad Eq.\ (3\text{-}24)\ *$$

Notice that the voltage gain is maximum when R_L is infinite and R_g is zero. Under these conditions the maximum

$$VG = \frac{r_{21}}{r_{11}} \qquad Eq.\ (3\text{-}25)\ *$$

For the typical point-contact transistor, the maximum $VG = \dfrac{24,000}{250}$ = 96. Assuming typical values of $R_L = 25,000$ ohms, and $R_g = 200$ ohms

$$VG = \frac{(24,000)\ (25,000)}{(25,000 + 12,000)\ (200 + 250) - 100\ (24,000)} = 42.1$$

For the typical junction transistor, the maximum $VG = \dfrac{1,900,000}{550} =$ 3,450. Assuming typical values $R_L = 1$ megohm, and $R_g = 200$ ohms

$$VG = \frac{(2,000,000)\ (1,000,000)}{(1,000,000 + 2,000,000)\ (200 + 550) - 550\ (1,900,000)} = 1,665$$

A comparison of the maximum and operating gains of the typical point-contact and junction transistors indicates that the junction is capable of furnishing much larger voltage gains. This explains why the junction transistor is invariably used in audio amplifier circuits.

The power gain (PG) of the transistor can be calculated from the product of the current gain and the voltage gain or found directly from the ratio of output power to input power.

$$PG = a(VG)$$

The theoretical maximum power gain is the maximum current gain and the maximum voltage gain. However, the condition for maximum current gain is $R_L = 0$, and the condition for maximum voltage gain is $R_L = $ infinity. Since these conditions are in opposition, the problem of finding the maximum power gain involves matching the input and output resistances of the transistor. The maximum power gain is obtained when the internal resistance of the signal generator is equal to the input resistance of the transistor, and the load resistance is equal to the output resistance of the transistor, that is $R_g = r_i$ and $R_L = r_o$. When these conditions are simultaneously satisfied, the transistor is *image impedance matched*.

Input and Output Impedance Matching. Equations 3-13 and 3-21 indicate that the input resistance is affected by the load resistance and, conversely, the output resistance depends on the generator internal resistance. Thus, starting with a given load resistance, if the generator resistance is changed to match the input resistance, the output resistance of the transistor changes, thus requiring a change in load resistance, and so on. In the following analysis, the proper values of generator and load resistance which satisfy both the input and output matching conditions at the same time are determined. Let r_1 equal the proper value of input resistance and generator resistance. Let r_2 equal the image matched value for the transistor output resistance and the load resistance. Then: $r_1 = R_g = r_i$ and $r_2 = R_L = r_o$.

Substituting for R_L and r_i in equation 3-13

$$r_1 = r_i = R_g = r_{11} - \left(\frac{r_{12}r_{21}}{r_2 + r_{22}} \right) \qquad Eq. \ (3\text{-}26)$$

Solving in terms of $r_{12}r_{21}$

$$(r_1 - r_{11})(r_2 + r_{22}) = -r_{12}r_{21} \qquad Eq. \ (3\text{-}27)$$

Substituting for R_g and r_o in equation 3-21

$$r_2 = r_o = R_L = r_{22} - \frac{r_{12}r_{21}}{r_1 + r_{11}} \qquad Eq. \ (3\text{-}28)$$

Again solving in terms of $r_{12}r_{21}$

$$(r_2 - r_{22})(r_1 + r_{11}) = -r_{12}r_{21} \qquad Eq. \ (3\text{-}29)$$

Equating equations 3-27 and 3-29

$$(r_1 - r_{11})(r_2 + r_{22}) = (r_2 - r_{22})(r_1 + r_{11}) \qquad Eq. \ (3\text{-}30)$$

Cross multiplying and cancelling equal terms,

$$r_1r_2 - r_2r_{11} + r_1r_{22} - r_{11}r_{22} = r_1r_2 - r_1r_{22} + r_2r_{11} - r_{11}r_{22}$$
$$2r_1r_{22} = 2r_2r_{11} \qquad\qquad Eq. \ (3\text{-}31)$$

or
$$\frac{r_2}{r_1} = \frac{r_{22}}{r_{11}} \qquad Eq.\ (3\text{-}32)$$

This latter equation indicates that matching the input and output resistances for maximum power gain requires their values to be in the same ratio as the open-circuit characteristics of the transistor.

The absolute value of the generator internal resistance and its matched input resistance in terms of transistor open-circuit parameters can now be determined. Substituting the equality $r_2 = \frac{r_1 r_{22}}{r_{11}}$ into equation 3-26,

$$r_1 = r_{11} - \left(\frac{r_{12}r_{21}}{\frac{r_1 r_{22}}{r_{11}} + r_{22}}\right) = r_{11} - \left(\frac{r_{12}r_{21}r_{11}}{r_{22}(r_1 + r_{11})}\right) \qquad Eq.\ (3\text{-}33)$$

$$(r_1 - r_{11})(r_1 + r_{11}) = -\left(\frac{r_{12}r_{21}r_{11}}{r_{22}}\right) = r_1{}^2 - r_{11}{}^2 \qquad Eq.\ (3\text{-}34)$$

$$r_1{}^2 = r_{11}{}^2 - \left(\frac{r_{12}r_{21}r_{11}}{r_{22}}\right) \qquad Eq.\ (3\text{-}35)$$

$$r_1 = \sqrt{r_{11}{}^2 - \left(\frac{r_{12}r_{21}r_{11}}{r_{22}}\right)} = \sqrt{\frac{r_{11}}{r_{22}}(r_{11}r_{22} - r_{12}r_{21})} \qquad Eq.\ (3\text{-}36) \bullet$$

In terms of the stability factor, $\delta = \frac{r_{12}r_{21}}{r_{11}r_{22}}$, which will be defined later in the chapter, the input image resistance

$$r_1 = \sqrt{r_{11}{}^2\left(\frac{r_{11}r_{22}}{r_{11}r_{22}}\right) - \left(\frac{r_{12}r_{21}}{r_{11}r_{22}}\right)} = r_{11}\sqrt{1 - \delta} \quad Eq.\ (3\text{-}37) \bullet$$

For the typical point-contact transistor previously considered, when $r_{11} = 250$ ohms, $r_{12} = 100$ ohms, $r_{21} = 24{,}000$ ohms, and $r_{22} = 12{,}000$ ohms, the numerical value of r_1 is

$$r_1 = \sqrt{\frac{250}{12{,}000}\Big[250\,(12{,}000) - 100\,(24{,}000)\Big]} \quad = 112 \text{ ohms}$$

For the typical junction transistor, when $r_{11} = 550$ ohms, $r_{12} = 500$ ohms, $r_{21} = 1{,}900{,}000$ ohms, and $r_{22} = 2{,}000{,}000$ ohms,

$$r_1 = \sqrt{\frac{550}{2{,}000{,}000}\Big[550\,(2{,}000{,}000) - 500\,(1{,}900{,}000)\Big]} \quad = 203 \text{ ohms}$$

The output image resistance of a transistor can be determined in a similar fashion from the ratio

$$\frac{r_2}{r_1} = \frac{r_{22}}{r_{11}}; \quad r_1 = \frac{r_2 r_{11}}{r_{22}}$$

Substituting this equality into equation 3-28

$$r_2 = r_{22} - \left(\frac{r_{12}r_{22}}{\frac{r_2 r_{11}}{r_{22}} + r_{11}}\right) = r_{22} - \left(\frac{r_{12}r_{21}r_{22}}{r_{11}(r_2 + r_{22})}\right) \qquad Eq.\ (3\text{-}38)$$

$$(r_2 - r_{22})(r_2 + r_{22}) = -\left(\frac{r_{12}r_{21}r_{22}}{r_{11}}\right) = r_2{}^2 - r_{22}{}^2 \qquad Eq. \ (3\text{-}39A)$$

$$r_2{}^2 = r_{22}{}^2 - \left(\frac{r_{12}r_{21}r_{22}}{r_{11}}\right) \qquad Eq. \ (3\text{-}39B)$$

$$r_2 = \sqrt{r_{22}{}^2 - \left(\frac{r_{12}r_{21}r_{22}}{r_{11}}\right)} = \sqrt{\frac{r_{22}}{r_{11}}(r_{11}r_{22} - r_{12}r_{21})} \qquad Eq. \ (3\text{-}40) \ \bullet$$

In terms of the stability factor $\delta = \dfrac{r_{12}r_{21}}{r_{11}r_{22}}$; the output image resistance

$$r_2 = \sqrt{r_{22}{}^2 \left(\frac{r_{11}r_{22}}{r_{11}r_{22}} - \frac{r_{12}r_{21}}{r_{11}r_{22}}\right)} = r_{22}\sqrt{1 - \delta} \qquad Eq. \ (3\text{-}41) \ \bullet$$

For the typical point-contact transistor,

$$r_2 = \sqrt{\frac{12,000}{250}\Big[250\,(12,000) - 100\,(24,000)\Big]} \quad = 5,370 \text{ ohms}$$

For the typical junction transistor

$$r_2 = \sqrt{\frac{2,000,000}{550}\Big[550\,(2,000,000) - 550\,(1,900,000)\Big]} \quad = 740,000 \text{ ohms}$$

These values may be checked on the R_L vs r_i and R_g vs r_o characteristics plotted for these typical transistors in Figs. 3-11, 3-12, 3-14, and 3-15.

Negative Resistance and Transistor Stability. Consider the general expression for input resistance

$$r_i = r_{11} - \left(\frac{r_{12}r_{21}}{r_{22} + R_L}\right) \qquad Eq. \ (3\text{-}13)$$

It is evident that the input resistance can have a negative value. The input resistance r_i is positive as long as r_{11} is greater than $\dfrac{r_{12}r_{21}}{r_{22} + R_L}$. This condition is most difficult to attain when the output is shorted, namely when $R_L = 0$. For the transistor to be stable under this condition, $r_{11}r_{22}$ must be greater than $r_{12}r_{21}$. The stability factor is the ratio of $r_{12}r_{21}$ to $r_{11}r_{22}$. The stability factor $\delta = \dfrac{r_{12}r_{21}}{r_{11}r_{22}}$ must be less than unity for short-circuit stability. Substituting the equivalent transistor parameters for the grounded base connection into the stability equation, the following relationship is obtained:

$r_{11}r_{22} > r_{12}r_{21}$ becomes $\ (r_c + r_b)(r_e + r_b) > r_b(r_m + r_b) \qquad Eq. \ (3\text{-}42)$

Expanding equation *3-42,*

$$r_c r_e + r_c r_b + r_b r_e + r_b{}^2 > r_b r_m + r_b{}^2$$

Dividing through by r_b

$$r_c + r_e + \frac{r_e r_c}{r_b} > r_m \qquad Eq. \ (3\text{-}43)$$

This equation emphasizes the importance of the backward transfer resistance r_b, since when $r_b = 0$, the transistor must have a positive input resistance.

On the other hand, if the value of r_b is increased by adding external resistance, it is possible to reach a condition where a normally positive input resistance becomes negative. Notice, however, that increasing the total base resistance eventually causes the input resistance to become negative only if $r_e + r_c$ is less than r_m. In the case of the junction transistor, r_c is always greater than r_m, and increasing the base resistance cannot produce a negative input resistance.

The conditions for negative output resistance are obtained similarly. In the general output resistance equation,

$$r_o = r_{22} - \left(\frac{r_{12}r_{21}}{R_g + r_{11}} \right) \qquad Eq. \ (3\text{-}21)$$

the output resistance r_o is positive provided that r_{22} is greater than

$$\frac{r_{12}r_{21}}{R_g + r_{11}}$$

This condition for stability is most difficult to meet when the generator resistance is equal to zero. For the transistor to be stable under this condition, $r_{11}r_{22}$ again must be greater than $r_{12}r_{21}$. The same stability factor and equations then exist for both the input and output resistances. It is evident, then, that one method of fabricating a transistor oscillator is by adding sufficient resistance to the base arm. Typical circuits incorporating this principle will be considered in Chapter 6.

Power Gain. Before determining the power gain included in transistor circuits, some definitions must be considered. Figure 3-16 illustrates a signal generator E_g with an internal resistance R_g feeding into a load R_L. The total power delivered by the generator $P_o = \dfrac{E_g^2}{R_g + R_L}$; the power dissipated in the load $P_L = \dfrac{E_L^2}{R_L}$. Since $E_L = \dfrac{E_g R_L}{R_g + R_L}$ then $P_L = \dfrac{E_g^2 R_L}{(R_g + R_L)^2}$

By using conventional calculus methods for determining conditions for maximum power, it is found that the load power is maximum when $R_g = R_L$. Under this condition the *power available from the generator*

$$P_a = \frac{E_g^2 R_g}{(2R_g)^2} = \frac{E_g^2}{4R_g}$$

The operating gain, G, of a network is defined as the ratio of the power dissipated in the load to the power available from the generator.

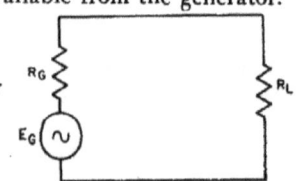

Fig. 3-16. Simplified transistor equivalent circuit for analysis of power gain.

For the general transistor circuit of Fig. 3-9

$$E_2 = i_2 R_L = \frac{E_g r_{21} R_L}{(R_g + r_{11}) (R_L + r_{22}) - r_{12} r_{21}} \qquad Eq.\ (3\text{-}44)$$

The power dissipated in the load

$$P_L = \frac{E_2^2}{R_L} = \left[\frac{r_{21}}{(R_g + r_{11}) (R_L + r_{22}) - r_{12} r_{21}} \right]^2 E_g^2 R_L \quad Eq.\ (3\text{-}45)$$

The operating gain

$$G = \frac{P_L}{P_a} = 4 R_g R_L \left[\frac{r_{21}}{(R_g + r_{11}) (R_L + r_{22}) - r_{12} r_{21}} \right]^2 \qquad Eq.\ (3\text{-}46)\ *$$

The available gain, AG, of a network is defined as the ratio of the power dissipated in the load to the power available from the generator when the load is matched to the output resistance. When $R_L = r_o = r_{22} - \left(\dfrac{r_{12} r_{21}}{R_g + r_{11}} \right)$, then $AG = \dfrac{P_L}{P_a}$

Substituting in equation 3-46, the available gain

$$AG = \frac{4 R_g \left(r_{22} - \dfrac{r_{12} r_{21}}{r_{11} R_g} \right) r_{21}^2}{\left[(R_g + r_{11}) (r_{22} - \dfrac{r_{12} r_{21}}{r_{11} + R_g} + r_{22}) - r_{12} r_{21} \right]^2} \qquad Eq.\ (3\text{-}47)$$

$$AG = \frac{4 R_g \left(r_{22} - \dfrac{r_{12} r_{21}}{r_{11} + R_g} \right) r_{21}^2}{4 (R_g + r_{11})^2 \left[r_{22} - \left(\dfrac{r_{12} r_{21}}{r_{11} + R_g} \right) \right]^2} \qquad Eq.\ (3\text{-}48)$$

$$AG = \frac{R_g r_{21}^2}{(R_g + r_{11})^2 \left[r_{22} - \left(\dfrac{r_{12} r_{21}}{r_{11} + R_g} \right) \right]} = \frac{R_g r_{21}^2}{(R_g + r_{11})^2 r_o} \qquad Eq.\ (3\text{-}49)\ *$$

The maximum available gain, MAG, of a network is defined as the ratio of the power dissipated in the load to the power available from the generator when the generator internal resistance is matched to the input transistor resistance, and when the load resistance is matched to the transistor output resistance. In order to solve for the maximum power gain in terms of the open-circuit parameters, the image-matched input and output resistances, previously determined, are substituted in the operating gain equation 3-46. Then, the maximum available gain,

$$MAG = \frac{4 r_1 r_2 r_{21}^2}{\left[(r_1 + r_{11}) (r_2 + r_{22}) - r_{12} r_{21} \right]^2} \qquad Eq.\ (3\text{-}50)$$

where $\qquad\qquad r_1 = r_{11} \sqrt{1 - \delta} \qquad\qquad\qquad Eq.\ (3\text{-}37)$

and $\qquad\qquad r_2 = r_{22} \sqrt{1 - \delta} \qquad\qquad\qquad Eq.\ (3\text{-}41)$

Substituting equations 3-37 and 3-41 in equation 3-50, for r_1 and r_2,

$$MAG = \frac{4 (r_{11} \sqrt{1 - \delta}) (r_{22} \sqrt{1 - \delta}) r_{21}^2}{\left[(r_{11} \sqrt{1 - \delta} + r_{11}) (r_{22} \sqrt{1 - \delta} + r_{22}) - r_{12} r_{21} \right]^2} \qquad Eq.\ (3\text{-}51)$$

which is equal to

$$\frac{4r_{11}r_{22}r_{21}^2 (1-\delta)}{\left[r_{11}r_{22}(1+\sqrt{1-\delta})^2 - r_{12}r_{21}\right]^2} = \frac{4r_{11}r_{22}r_{21}^2 (1-\delta)}{r_{11}^2 r_{22}^2 \left[(1+\sqrt{1-\delta})^2 - \left(\frac{r_{12}r_{21}}{r_{11}r_{22}}\right)^2\right]}$$

Eq. (3-52)

$$MAG = \frac{4r^2_{21} (1-\delta)}{r_{11}r_{22} \left[(1+\sqrt{1-\delta})^2 - \delta\right]^2} = \frac{4r^2_{21} (1-\delta)}{r_{11}r_{22} \left[1 + 2\sqrt{1-\delta} + 1 - \delta - \delta\right]^2}$$

Eq. (3-53)

from which derives

$$\frac{4r^2_{21} (1-\delta)}{4r_{11}r_{22} \left[1 + \sqrt{1-\delta} - \delta\right]^2} = \frac{r^2_{21} (\sqrt{1-\delta})^2}{r_{11}r_{22} (\sqrt{1-\delta})^2 (1 + \sqrt{1-\delta})^2}$$

Eq. (3-54)

$$MAG = \frac{r^2_{21}}{r_{11}r_{22} (1 + \sqrt{1-\delta})^2}$$ 　　Eq. (3-55) *

For the typical point-contact transistor, when $r_{11} = 250$ ohms, $r_{12} = 100$ ohms, $r_{21} = 24,000$ ohms, $r_{22} = 12,000$ ohms, and when assuming $R_g = 50$ ohms and $R_L = 8,000$ ohms, the operating gain G, becomes

$$G = \frac{4R_g R_L r^2_{21}}{\left[(R_g + r_{11}) (R_L + r_{22}) - r_{12}r_{21}\right]^2}$$

$$G = \frac{4 (50) (8,000) (24,000)^2}{\left[(50 + 250) (8,000 + 12,000) - 100 (24,000)\right]^2} = 71.3$$

The available gain, $AG = \dfrac{R_g r^2_{21}}{\left[r_{22} - \left(\dfrac{r_{12}r_{21}}{r_{11} + R_g}\right)\right](r_{11} + R_g)^2}$

$$AG = \frac{(50) (24,000)^2}{\left[12,000 - \dfrac{100 (24,000)}{250 + 50}\right](250 + 50)^2} = 80$$

The maximum available gain $MAG = \dfrac{r^2_{21}}{r_{11}r_{22} (1 + \sqrt{1-\delta})^2}$

$$MAG = \frac{(24,000)^2}{250 (12,000) \left(1 + \sqrt{1 - \dfrac{100 (24,000)}{250 (12,000)}}\right)^2} = 92$$

Notice that the stability factor, $\delta = \dfrac{r_{12}r_{21}}{r_{11}r_{22}}$, is $\dfrac{100 (24,000)}{250 (12,000)} = 0.8$. If the stability factor is greater than one, the numerical value of the quantity $\sqrt{1-\delta}$ must be negative, which indicates an unstable condition. For the typical junction transistor in which $r_{11} = 550$ ohms, $r_{12} = 500$ ohms, $r_{21} = 1,900,000$ ohms, and $r_{22} = 2,000,000$ ohms; when assuming $R_g = 100$ ohms and $R_L = 1,000,000$ ohms, the operating gain

$$G = \frac{4R_g R_L r^2{}_{21}}{\left[(R_g + r_{11})\,(R_L + r_{22}) - r_{12}r_{21}\right]^2}$$

$$G = \frac{4\,(100)\,(1{,}000{,}000)\,(1{,}900{,}000)^2}{\left[(100 + 550)\,(1{,}000{,}000 + 2{,}000{,}000) - 500\,(1{,}900{,}000)\right]^2} = 1{,}444$$

The available gain $AG = \dfrac{R_g r^2{}_{21}}{\left[r_{22} - \left(\dfrac{r_{12}r_{21}}{r_{11} + R_g}\right)\right](r_{11} + R_g)^2}$

$$AG = \frac{100\,(1{,}900{,}000)^2}{\left[2{,}000{,}000 - \dfrac{500\,(1{,}900{,}000)}{550 + 100}\right](550 + 100)^2} = 1{,}583$$

From equation (3-55)

$$MAG = \frac{(1{,}900{,}000)^2}{500\,(2{,}000{,}000)\left(1 + \sqrt{1 - \dfrac{500\,(1{,}900{,}000)}{550\,(2{,}000{,}000)}}\,\right)^2} = 2{,}400$$

Junction Capacitance

Before the actual transistor circuit is considered, some additional and important characteristics must be defined. In Chapter 2, the collector junction capacitance was mentioned in connection with transistor high-frequency characteristics. In the equivalent network, this parameter acts in parallel with the collector resistance. The value of collector junction capacitance C_c varies in units of the same type, but for a typical junction transistor is approximately 10 $\mu\mu$f. The value of capacitance is primarily a function of the junction area, although it also depends on the width of the junction layer and the resistivity of the base layer.

Zener Voltage

If the reverse voltage applied across a P-N junction is gradually increased, a point is reached where the potential is high enough to break down covalent bonds and cause current flow. This voltage is called *Zener voltage*. In transistor application the Zener voltage has the same design importance as the inverse voltage rating of a vacuum tube, since it defines the maximum reverse voltage which can be applied to a junction without excessive current flow. The Zener potential for a transistor junction can be increased by widening the space charge layer, or by forming the junction so that the transition from N region to P region is a gradual process. In germanium, the Zener voltage field is around 2 x 10^5 volts/centimeter. A transistor junction with a Zener potential of 70 volts would therefore have a space charge layer of

$$\frac{70}{2 \times 10^5} = 35 \times 10^{-5} \text{ centimeters.}$$

Saturation Current, I_{co}

Another important transistor characteristic is the *saturation current* I_{co}. This is the collector current that flows when the emitter current is zero. In properly functioning transistors, I_{co} is in the vicinity of 10 microamperes; the value is considerably higher in defective transistors. The saturation current is composed of two components. The first is formed by thermally generated carriers which diffuse into the junction region. The second component is an ohmic characteristic which is caused by surface leakage across the space charge region, from local defects in the germanium, or from a combination of these two factors. The ohmic component may be separated from the true or thermally caused value by measuring the collector current at different values of collector voltage.

Chapter 4
GROUNDED EMITTER AND GROUNDED COLLECTOR TRANSISTORS

The design and servicing of the transistor circuit is more complicated than that of the vacuum tube, because transistor input and output circuits are never inherently independent of each other. This makes it difficult for a newcomer to get the "feel" of the transistor. In the long run, however, these same complex characteristics provide for a more flexible device, one capable of many circuit applications beyond the range of the vacuum tube.

This chapter deals with the extension of the four-terminal characteristics developed for the grounded base to encompass the two remaining connections, the grounded emitter, and the grounded collector; a comparison of the major features of the three basic connections; limitations of the transistor; and transistor testing methods.

Introduction

In the following analysis of transistor performance in the grounded emitter and grounded collector connections, the same typical point-contact and junction transistors discussed in Chapter 3 will be used for numerical examples. For the point-contact transistor in the grounded base connection, the parameters are:

$r_{12} = 100$ ohms $= r_b$
$r_{11} = 250$ ohms $= r_e + r_b$; then $r_e = 150$ ohms
$r_{21} = 24{,}000$ ohms $= r_m + r_b$; then $r_m = 23{,}900$ ohms
$r_{22} = 12{,}000$ ohms $= r_c + r_b$; then $r_c = 11{,}900$ ohms

For the junction transistor in the grounded base connection:

$r_{12} = 500$ ohms $= r_b$
$r_{11} = 550$ ohms $= r_e + r_b$; then $r_e = 50$ ohms
$r_{21} = 1{,}900{,}000$ ohms $= r_m + r_b$; then $r_m = 1{,}899{,}500$ ohms
$r_{22} = 2{,}000{,}000$ ohms $= r_c + r_b$; then $r_c = 1{,}999{,}500$ ohms

Notice that since r_m and r_c are so much greater in value than r_b, particularly in the case of the junction transistor, for all practical purposes $r_{21} = r_m$ and $r_{22} = r_c$.

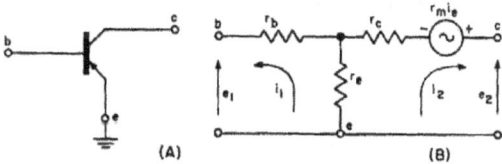

Fig. 4-1. (A) The grounded emitter connection. (B) Equivalent active "T" for grounded emitter connection.

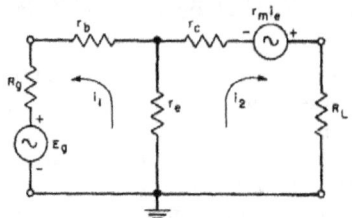

Fig. 4-2. Operating circuit, grounded emitter connection.

The Grounded Emitter Connection

Equivalent Operating Circuit. The grounded emitter connection is illustrated in Fig. 4-1 (A). In this case the input connection is made between the base and emitter electrodes (conventionally the emitter is shown schematically as an arrowhead resting on the base), and the output is taken between the collector and the emitter. Thus, in this case, the emitter is the common electrode. Figure 4-1 (B) illustrates the equivalent active "T" circuit for the grounded emitter connection.

Figure 4-2 is the complete operating circuit of the grounded emitter connection. Notice that although the negative side of the signal generator is grounded, the polarity of the signal in this connection is reversed with respect to the emitter and base terminals shown in the grounded base connection of Fig. 3-9. Since this effective reversal of input leads is the only physical difference between the two connections, the grounded emitter, unlike the grounded base, produces a phase inversion of the input signal.

Circuit Parameters. The general open-circuit characteristics derived for the grounded base connection apply equally well to the grounded emitter and grounded base connections, since the characteristics were determined on the basis of a sealed box. However, since the internal parameters of the transistor have been rearranged, the values of the general characteristics are different. It is necessary then, to evaluate the open-circuit characteristics r_{11}, r_{12}, r_{21}, and r_{22} in terms of the transistor internal parameters r_e, r_b, r_c, and r_m. The same basic measuring circuits, illustrated in Figs. 3-8, may be used to determine the four-terminal parameter for the grounded emitter connection:

A: $r_{11} = e_1/i_1$ when $i_2 = 0$, $r_{11} = r_e + r_b$ Eq. (4-1)
B: $r_{21} = e_2/i_1$ when $i_2 = 0$, $r_{21} = r_e - r_m$ Eq. (4-2)
C: $r_{12} = e_1/i_2$ when $i_1 = 0$, $r_{12} = r_e$ Eq. (4-3)
D: $r_{22} = e_2/i_2$ when $i_1 = 0$, $r_{22} = r_e + r_c - r_m$ Eq. (4-4)

These grounded-emitter relationships are derived as follows:

A. Using Fig. 4-1 (B), the input loop equation on the basis of Kirchoff's law is:

$$e_1 = i_1 (r_e + r_b) + i_2 r_e$$

when
$$i_2 = 0, \quad e_1 = i_1 (r_e + r_b)$$

then
$$r_{11} = \frac{e_1}{i_1} = \frac{i_1 (r_e + r_b)}{i_1} = r_e + r_b$$

B. For the same input loop equation, when $i_1 = 0$
$$e_1 = i_2 r_e;$$

then
$$r_{12} = \frac{e_1}{i_2} = \frac{i_2 r_e}{i_2} = r_e$$

C. The output loop equation for Fig. 4-1 (B) on the basis of Kirchoff's law is:
$$e_2 - r_m i_e = i_1 r_e + i_2 (r_e + r_c)$$

Also
$$i_e = - (i_1 + i_2)$$

Substituting for i_e,
$$e_2 + r_m (i_1 + i_2) = i_1 r_e + i_2 (r_e + r_c)$$
$$e_2 = i_1 (r_e - r_m) + i_2 (r_e + r_c - r_m)$$

when
$$i_2 = 0, \quad e_2 = i_1 (r_e - r_m)$$

then
$$r_{21} = \frac{e_2}{i_1} = \frac{i_1 (r_e - r_m)}{i_1} = r_e - r_m$$

D. Using the same equations as in C above, when
$$i_1 = 0, \quad e_2 = i_2 (r_e + r_c - r_m)$$

then
$$r_{22} = \frac{e_2}{i_2} = \frac{i_2 (r_e + r_c - r_m)}{i_2}, = r_e + r_c - r_m$$

The open-circuit characteristics can now be numerically evaluated for the typical point-contact and junction transistors previously considered in Chapter 3. For the point-contact transistor in the grounded emitter connection:

$r_{11} = r_e + r_b = 150 + 100 = 250$ ohms
$r_{12} = r_e = 150$ ohms
$r_{21} = r_e - r_m = 150 - 23,900 = -23,750$ ohms
$r_{22} = r_e + r_c - r_m = 150 + 11,900 - 23,900 = -11,850$ ohms

For the junction transistor in the grounded emitter connection:

$r_{11} = r_e + r_b = 50 + 500 = 550$ ohms
$r_{12} = r_e = 50$ ohms
$r_{21} = r_e - r_m = 50 - 1,899,500 = -1,899,450$ ohms
$r_{22} = r_e + r_c - r_m = 50 + 1,999,500 - 1,899,500 = 100,050$

Because of the large values of r_m and r_c with respect to r_e, r_{21} in the practical case can be approximated by $-r_m$, and r_{22} by $(r_c - r_m)$. The emitter resistance, $r_e = r_{12}$, is the feedback resistance and is equivalent to $r_b = r_{12}$ in the grounded base connection. Notice, however, that since there is phase inversion in the grounded emitter connection, r_e produces degenerative (negative) feedback, rather than regenerative (positive) feedback. The degenerative effect of the output current through r_e is similar to the degenerative action of an unbypassed cathode resistor in a grounded cathode vacuum tube.

Current Gain in the Grounded Emitter Connection. The current gain in terms of the general four-terminal parameters was defined by equation *3-8 as:*

$$a = \frac{r_{21}}{R_L + r_{22}} \qquad\qquad Eq. \ (4\text{-}5)$$

In terms of the transistor parameters in the grounded emitter connection now being considered, the current gain is

$$a = \frac{r_e - r_m}{R_L + r_e + r_c - r_m} \qquad\qquad Eq. \ (4\text{-}6) *$$

In the case of the grounded-emitter point-contact transistor, r_{21} and r_{22} are both negative. The value of the load resistor, R_L, determines whether the current gain is positive or negative. If R_L is less than the absolute value of $-r_{22}$, a is positive; if R_L is greater than the absolute value of $-r_{22}$, a is negative. A negative value of current gain indicates simply that the input current is inverted in phase. This is normal in the grounded emitter connection. Theoretically, an infinite current gain is attained when $R_L = -r_{22}$. The current gain of a typical point-contact transistor with a load $R_L = 15,000$ ohms is

$$a = \frac{-23,750}{15,000 - 11,850} = -7.54$$

(Equation 3-8A for maximum current gain, $a_0 = \dfrac{r_{21}}{r_{22}}$, does not apply

in this connection, since it is found that the point-contact transistor is unstable when R_L is less than $-r_{22}$.) Notice that the current gain becomes very large for values of R_L slightly larger than $-r_{22}$. For example, if

$$R_L = 12,500 \text{ ohms, } a = \frac{-23,750}{12,500 - 11,850} = -36.6$$

The current gain in the junction transistor is always negative in the grounded emitter connection, since r_{22} is always positive. The current gain for the typical grounded-emitter junction transistor with a load

$$R_L = 100,000 \text{ ohms, is } a = \frac{-1,899,450}{100,000 + 100,050} = -9.5$$

The maximum current gain, as in the case of the grounded base connection, is:

$$a_0 = \frac{r_{21}}{r_{22}} = \frac{-1,899,450}{100,050} = -19.0$$

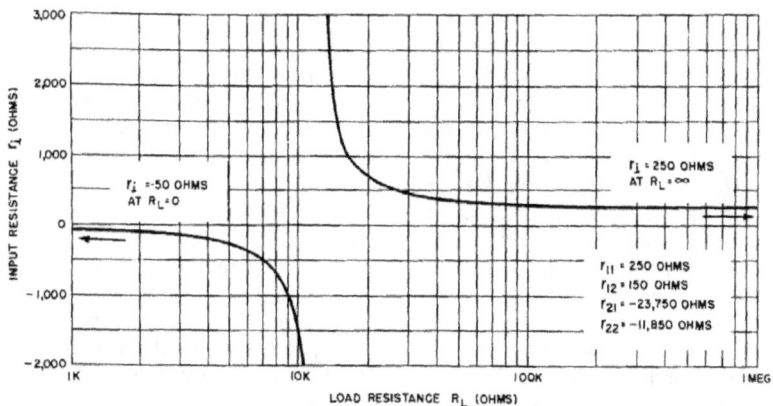

Fig. 4-3. Input resistance vs load resistance for typical point-contact transistor
(grounded emitter).

Input Resistance r_i for the Grounded Emitter Connection. The input resistance was defined in equation *3-13* in terms of the general open-circuit parameters as:

$$r_i = r_{11} - \left(\frac{r_{12}r_{21}}{r_{22} + R_L} \right)$$

The input resistance in terms of the transistor parameters in the grounded emitter connection becomes:

$$r_i = r_e + r_b - \left(\frac{r_e (r_e - r_m)}{r_e + r_c - r_m + R_L} \right) \qquad Eq. \ (4\text{-}7) \ *$$

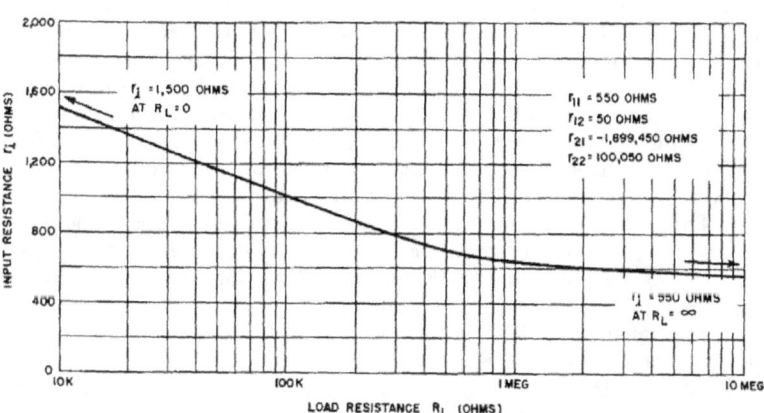

Fig. 4-4. Input resistance vs load resistance for typical junction transistor (grounded emitter).

The effect of the value of the load resistance on the input resistance of typical transistors is illustrated in Figs. 4-3 and 4-4. The input resistance for the point-contact transistor starts at a value of -40 ohms for $R_L = 0$, and becomes more negative as the load resistance increases. When $R_L = -r_{22}$, the input resistance is infinite. As the load resistance increases beyond this point, the input resistance becomes positive, decreasing in value to the limiting condition $r_i = r_{11} = 250$ ohms when the output is open-circuited. Negative values of input resistance indicate circuit instability; consequently, the point-contact transistor can be used as an oscillator in the region where R_L is less than $-r_{22}$. Circuits of this type are called "collector-controlled oscillators."

The input resistance of the junction transistor is always positive. In the typical transistor considered, the input resistance decreases from a value of 1,500 ohms at $R_L = 0$, to 550 ohms for an infinite load.

Output Resistance r_o for the Grounded Emitter Connection. The output resistance was defined by equation *3-21* in terms of the general four-terminal parameters as:

$$r_o = r_{22} - \left(\frac{r_{12}r_{21}}{R_g + r_{11}}\right)$$

The output resistance in terms of the internal transistor parameters in the grounded emitter connection becomes:

$$r_o = r_e + r_c - r_m - \left(\frac{r_e(r_e - r_m)}{R_g + r_e + r_b}\right) \qquad Eq. \ (4\text{-}8) \ *$$

The effect of the value of the signal generator resistance is illustrated for the point-contact and junction transistors in Figs. 4-5 and 4-6, respectively. Notice that the output resistance of the point-contact type is positive at $R_g = 0$, and decreases rapidly to zero when R_g is slightly less than 50 ohms. As R_g is increased further, r_o becomes negative and gradually approaches the limiting condition, when R_g is infinite, $r_i = r_{22} = -11,850$ ohms. Thus, the point-contact transistor can have a negative output resistance over a large range of generator resistance values, and this characteristic can be used in transistor oscillator design. Circuits of this type are called "base-controlled oscillators."

The output resistance of the junction transistor is always positive, and for the typical type considered, r_o gradually decreases from approximately 273,000 ohms to 100,000 ohms as R_g is increased from zero to infinity.

The range in which both the output and input resistances of the point-contact transistor are positive can be increased by adding external resistance in the emitter arm. This increases the effective value of $r_e = r_{12}$. Notice that if enough external resistance is added so that the effective emitter resistance $r_e + R_L$ is equal to or greater than

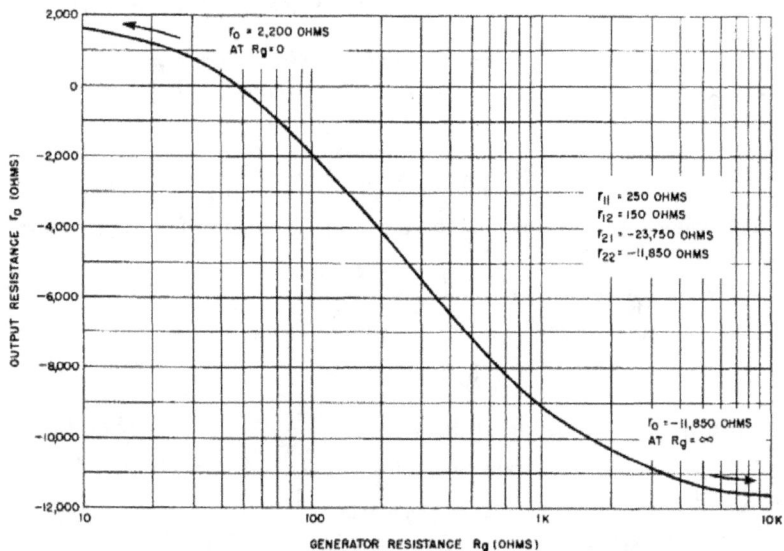

Fig. 4-5. Output resistance vs generator resistance for typical point-contact transistor (grounded emitter).

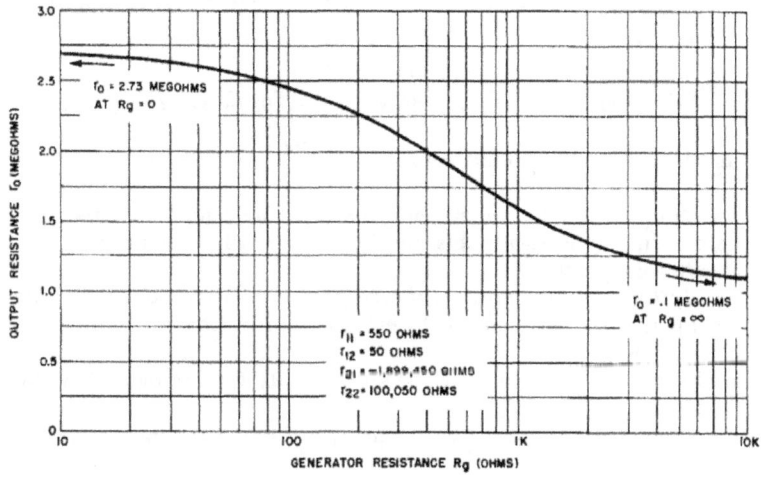

Fig. 4-6. Output resistance vs generator resistance for typical junction transistor (grounded emitter).

$- (r_c - r_m)$, the input and output resistance is positive. This stabilizing effect of adding resistance to the emitter load is frequently used in transistor circuit applications.

Voltage Gain VG in the Grounded Emitter Connection. The voltage gain was defined by equation *3-24* in terms of the general four-terminal parameters as:

$$VG = \frac{r_{21}R_L}{(R_L + r_{22})(R_g + r_{11}) - r_{12}r_{21}}$$

The voltage gain in terms of the transistor parameters for the grounded emitter connection becomes:

$$VG = \frac{(r_e - r_m)R_L}{(R_L + r_e + r_c - r_m)(R_g + r_e + r_b) - r_e(r_e - r_m)} \qquad Eq. \ (4\text{-}9) \ *$$

The maximum voltage gain defined by equation *3-25* is:

$$\text{Max. VG} = \frac{r_{21}}{r_{11}},$$

which in this case becomes $\quad \text{Max. VG} = \dfrac{r_e - r_m}{r_e + r_b} \qquad Eq. \ (4\text{-}9A)$

For the point-contact transistor, assuming $R_L = 30{,}000$ ohms and $R_g = 10$ ohms,

$$VG = \frac{-23{,}750\,(30{,}000)}{(30{,}000 - 11{,}850)(10 + 250) - (150)(-23{,}750)} = -86.5$$

and the maximum voltage gain is $\dfrac{-23{,}750}{250} = -95.0$

For the junction transistor, with $R_L = 100{,}000$ ohms and $R_g = 10$ ohms, the voltage gain is

$$VG = \frac{-1{,}899{,}450\,(100{,}000)}{(100{,}000 + 100{,}050)(10 + 550) - (50)(-1{,}899{,}450)} = -918$$

and the maximum voltage gain is $\dfrac{-1{,}899{,}450}{550} = -3{,}460$

Notice that the voltage gain in this connection, like the current gain, is negative. Again this merely indicates that the input voltage is inverted in phase.

Impedance Matching in the Grounded Emitter Connection. As in the analysis of the grounded base connection in Chapter 3, the stability factor $\delta = \dfrac{r_{12}r_{21}}{r_{11}r_{22}}$ must be less than unity for short-circuit stability.

The numerical value of δ for the typical point-contact transistor is

$\dfrac{150\,(-23{,}750)}{250\,(-11{,}850)} = 1.2$. This re-emphasizes the fact that the point-contact transistor in the grounded emitter connection is unstable when the output is short-circuited. The stability factor for the junction transistor is $\dfrac{50\,(-\,1,899,450)}{550\,(100,050)} = -\,1.73$, which confirms the fact that the junction transistor is short-circuit stable in the grounded emitter connection.

The input image resistance is defined in equation *3-37* as: $r_1 = r_{11}\sqrt{1-\delta}$. The input image resistance of the point-contact transistor is numerically equal to $250\sqrt{1-1.19} = 250\sqrt{-.19}$. Since the quantity under the square root sign is negative, r_1 is imaginary and cannot be built into conventional signal sources. For the typical junction transistor, r_1 is $550\sqrt{1-(-1.73)} = 908$ ohms.

The output image resistance defined by equation *3-41* as $r_2 = r_{22}\sqrt{1-\delta}$ also is imaginary for the point-contact transistor. For the junction transistor, r_2 is $100,050\sqrt{1-(-1.73)} = 165,000$ ohms.

Power Gain in the Grounded Emitter Connection. The numerical values of the voltage and current gains are always negative in the stable range of operation of the grounded emitter connection. The negative sign is merely a mathematical indication of the phase inversion of the amplified signal. Since the power gain is a function of the product of the voltage and current gains, its numerical value must be positive.

The operating gain is defined in equation *3-46* as

$$G = \frac{4R_g R_L r^2_{21}}{[(R_g + r_{11})\,(R_L + r_{22}) - r_{12}r_{21}]^{\,2}}$$

Note that since r_{21} and the bracketed quantity in the denominator are squared, the numerical value of this equation is always positive. It is certainly possible to obtain an apparently valid power gain in an unstable portion of the transistor characteristic if numerical values are haphazardly substituted. For example, evaluating the operating gain for the typical point-contact transistor when $R_L = 1,000$ ohms and $R_g = 10$ ohms,

$$G = \frac{4\,(10)\,(1000)\,(-23,750)^{\,2}}{[(10 + 250)\,(1000 - 11,850) - 150\,(23,750)]^{\,2}} = 41.4$$

However, Fig. 4-3 indicates that at a load of $R_L = 1,000$ ohms, the transistor is unstable and will oscillate. This does not mean that the grounded emitter connected transistor can oscillate and supply a power gain at the same time, but rather that the operating gain equation can only be applied conditionally. Without going too deeply into the mathematical concepts involved, equation *3-46* can only be applied when

$$(R_g + r_{11})\,(R_L + r_{22}) - r_{12}r_{21}$$

is greater than zero. When R_g and R_L equal zero, the worst possible case, the condition equation becomes $r_{11}r_{12} - r_{12}r_{21} > 0$. This is just another way of expressing the requirement that the stability factor, $\delta =$

$\dfrac{r_{12}r_{21}}{r_{11}r_{22}}$,must be less than unity. As a result, the operating gain is conditional when δ is greater than unity.

In the typical point-contact transistor under discussion, $\delta = 1.19$, the conditional equation is $(R_g + 250)(R_L - 11,850) - 150(-23,750) > 0$. A plot of this conditional characteristic is shown in Fig. 4-7. Any combination of generator resistance and load resistance in the stable region can be used, but the selection of operating values close to the conditional characteristic provides the greatest operating gain.

The following example illustrates the design of a grounded emitter circuit for maximum power gain when the stability factor is greater than one. Assume that the load R_L is fixed at 10,000 ohms for the typical point-contact transistor. Figure 4-7 indicates than any value of R_g less than 1,530 ohms will provide stable operation. Thus for $R_g = 100$ ohms

$$G = \frac{4(100)(10,000)(-23,750)^2}{\left[(100 + 250)(10,000 - 11,850) - 150(-23,750)\right]^2} = 267$$

If, however, $R_g = 1,450$ ohms were selected,

$$G = \frac{4(1450)(10,000)(-23,750)^2}{\left[(1450 + 250)(10,000 - 11,850) - 150(-23,750)\right]^2} = 195,000$$

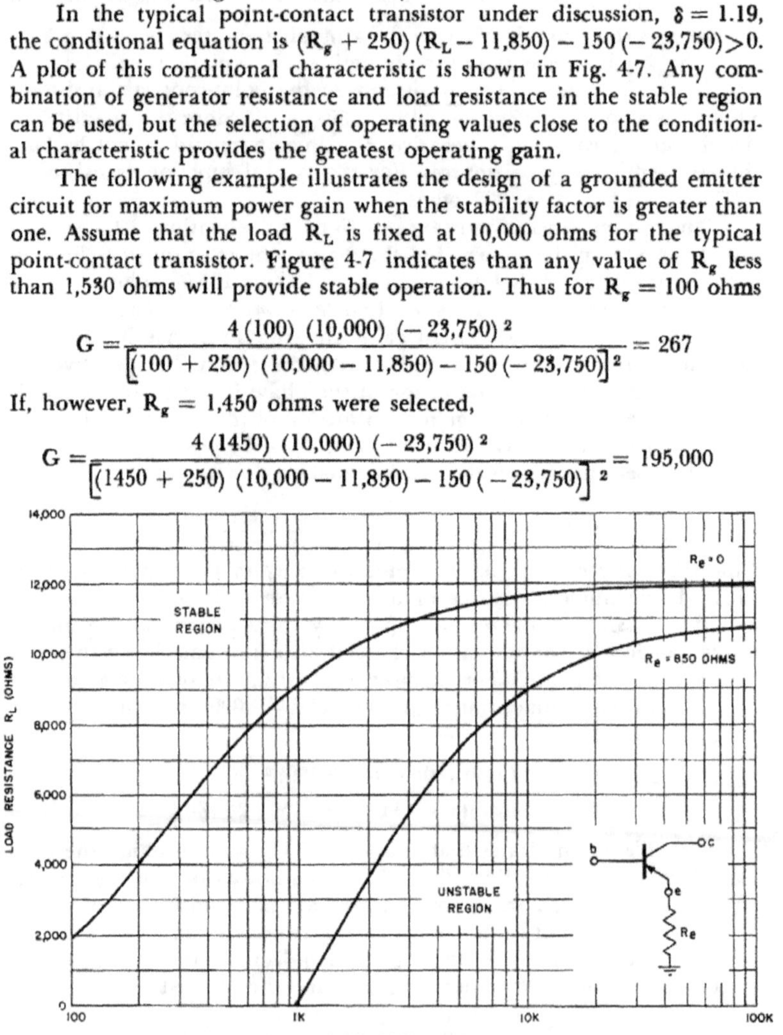

Fig. 4-7. Conditional stability characteristic (grounded emitter).

These examples prove that extremely high values of power gain can be attained by selection of $R_g R_L$ values close to the stability characteristic. In the practical case, however, the selected values must be sufficiently removed from the instability limit to avoid the introduction of circuit oscillation by normal parameter variations.

The grounded emitter connection can be stabilized by adding resistance in the emitter arm. As an example, assume that a resistor $R_e = 850$ ohms is added in series with the emitter. The four-terminal parameters then become:

$r_{11} = r_e + R_e + r_b = 150 + 850 + 100 = 1,100$ ohms

$r_{12} = r_e + R_e = 150 + 850 = 1,000$ ohms

$r_{21} = r_e + R_e - r_m = 150 + 850 - 23,900 = -22,900$ ohms

$r_{22} = r_e + R_e + r_c - r_m = 150 + 850 + 11,900 - 23,900 = -11,000$ ohms

Substituting these new values, the conditional equation becomes $(R_g + 1,100) (R_L - 11,000) - (1,000) (-22,900)$ and must be greater than zero. A plot of the modified conditional stability characteristic is shown in Fig. 4-7. Notice the extent to which the stability area has been increased. As before, the selection of values R_L and R_g located near the limiting line provide the greatest power gain.

The maximum available power gain defined by equation 3-55,

$$MAG = \frac{r^2_{21}}{r_{11}r_{12}(1 + \sqrt{1- \delta})^2}$$

can be applied to the grounded emitter connection provided that the stability factor is less than one. The numerical value of the maximum available gain for the typical junction transistor is:

$$MAG = \frac{(1,899,450)^2}{550(100,050)[1 + \sqrt{1 - (-1.73)}]^2} = 9,340$$

The Grounded Collector Connection

Equivalent Operating Circuit. The grounded collector connection is illustrated in Fig. 4-8 (A). In this connection, the input signal is connected between the base and collector electrodes, and the output is taken between the emitter and the common collector. The equivalent active "T" is illustrated in Fig. 4-8 (B).

The general four-terminal parameters can be measured in terms of the internal transistor parameters using the basic measuring circuits of Fig. 3-8. The four-terminal parameter equations for the grounded collector connection are:

A. $r_{11} = \dfrac{e_1}{i_1}$ when $i_2 = 0$, $r_{11} = r_b + r_c$ *Eq. (4-10)*

B. $r_{21} = \dfrac{e_2}{i_1}$ when $i_2 = 0$, $r_{21} = r_c$ *Eq. (4-11)*

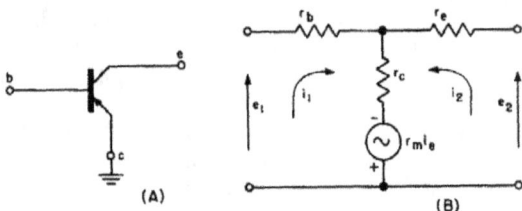

Fig. 4-8. (A) The grounded collector connection. (B) Equivalent active "T" for grounded collector connection.

C. $r_{12} = \dfrac{e_1}{i_2}$ when $i_1 = 0$, $r_{12} = r_c - r_m$ *Eq. (4-12)*

D. $r_{22} = \dfrac{e_2}{i_2}$ when $i_1 = 0$, $r_{22} = r_e + r_c - r_m$ *Eq. (4-13)*

These equalities are derived as follows:

A. Using Fig. 4-8 (B), the input loop equation on the basis of Kirchoff's law is:

$$e_1 + r_m i_e = i_1 (r_b + r_c) + i_2 r_c$$

For this connection $i_2 = i_e$

and $e_1 = i_1 (r_b + r_c) + i_2 (r_c - r_m)$

when $i_2 = 0$, $e_1 = i_1 (r_b + r_c)$,

and $r_{11} = \dfrac{e_1}{i_1} = \dfrac{i_1 (r_b + r_c)}{i_1} = r_b + r_c$

B. Using the same input loop equation, when $i_1 = 0$, $e_1 = i_2 (r_c - r_m)$,

and $r_{12} = \dfrac{e_1}{i_2} = \dfrac{i_2 (r_c - r_m)}{i_2} = r_c - r_m$

C. The output voltage loop equation is:

$$e_2 + r_m i_e = i_1 r_c + i_2 (r_e + r_c)$$

Since $i_2 = i_e$, $e_2 = i_1 r_c + i_2 (r_e + r_c - r_m)$

when $i_2 = 0$, $e_2 = i_1 r_c$

$$r_{21} = \dfrac{e_2}{i_1} = \dfrac{i_1 r_c}{i_1} = r_c$$

D. Using the same output loop equation,

when $i_1 = 0$, $e_2 = i_2 (r_e + r_c - r_m)$

$$r_{22} = \dfrac{e_2}{i_2} = \dfrac{i_2 (r_e + r_c - r_m)}{i_2} = r_e + r_c - r_m$$

The numerical values of the four-terminal parameters can now be determined for the typical point-contact transistor. The values are:

$r_{11} = r_b + r_c = 100 + 11,900 = 12,000$ ohms

$r_{12} = r_c - r_m = 11,900 - 23,900 = -12,000$ ohms

$r_{21} = r_c = 11,900$ ohms

$r_{22} = r_e + r_c - r_m = 150 + 11,900 - 23,900 = -11,850$ ohms

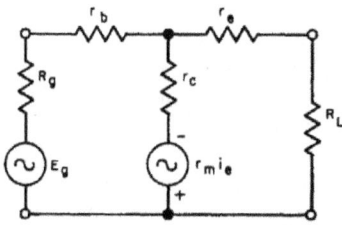

Fig. 4-9. Operating circuit, grounded collector connection.

The numerical values for the junction transistor are:

$r_{11} = r_b + r_c = 500 + 1,999,500 = 2,000,000$ ohms
$r_{12} = r_c - r_m = 1,999,500 - 1,899,500 = 100,000$ ohms
$r_{21} = r_c = 1,999,500$ ohms
$r_{22} = r_e + r_c - r_m = 50 + 1,999,500 - 1,899,500 = 100,050$ ohms

Because of the very low values of r_b and r_e compared to the quantities r_c and $(r_c - r_m)$, r_{11} is approximately equal to r_{21}, and r_{22} is approximately equal to r_{12}.

Figure 4-9 illustrates the operating circuit for the grounded collector connection. As in the analysis of the grounded emitter circuit, the performance characteristics for this connection can now be determined by straightforward substitution in the general four-terminal circuit equations.

Current Gain, a, of the Grounded Collector Connection. The current gain as defined in equation 3-8 is:

$$a = \frac{r_{21}}{R_L + r_{22}}$$

In terms of the internal transistor parameters in the grounded collector connection, the current gain becomes:

$$a = \frac{r_c}{R_L + r_e + r_c - r_m} \qquad Eq. \ (4\text{-}14)$$

The value of r_{22} is always negative in the case of the point-contact transistor. Therefore, the load resistor R_L must be larger than the absolute value of r_{22} for stable operation, and the equation for maximum current gain $a_o = \dfrac{r_{21}}{r_{22}}$ can only be applied to the junction transistor. Numerical values for the typical junction transistor when $R_L = 100,000$ ohms are

$$a = \frac{1,999,500}{100,000 + 100,050} = 10$$

and the maximum current gain

$$a_o = \frac{1,999,500}{100,050} = 20$$

For the point-contact transistor when $R_L = 15,000$ ohms,

$$a = \frac{11,900}{15,000 - 11,850} = 3.84$$

It would appear that as the load approaches the absolute value of r_{22}, extremely high current gains are attainable. For example, for the point-contact transistor when $R_L = 11,950$ ohms,

$$a = \frac{11,900}{11,950 - 11,850} = 119$$

In operating circuits, however, the grounded collector current gain is limited to the same order of magnitude as in the grounded emitter connection. This limitation is caused by the rapid increase in input resistance with an increase in current gain.

Input Resistance, r_i, in the Grounded Collector Circuit. The general input resistance is defined by equation *3-13*:

$$r_i = r_{11} - \left(\frac{r_{12} r_{21}}{R_L + r_{22}} \right)$$

In terms of grounded collector transistor parameters, the input resistance becomes

$$r_i = r_b + r_c - \left(\frac{(r_c - r_m) r_c}{(r_e + r_c - r_m) + R_L} \right) \qquad Eq.\ (4\text{-}15)$$

The variation of input resistance with load is illustrated for the point-contact and junction transistors in Figs. 4-10 and 4-11, respectively. The input resistance for the point-contact transistor is negative from $R_L = 0$ to $R_L = -r_{22} = 11,850$ ohms. Notice that when $R_L = -r_{22}$, the input resistance is infinite or open circuited. As R_L is increased further, the input resistance becomes positive, and gradually decreases to a limiting value of 12,000 ohms. The input resistance of the junction transistor increases from a value of approximately 500 ohms to the limiting value $r_i = r_{11} = 2,000,000$ ohms when the output is open circuited.

Output Resistance, r_o, for the Grounded Collector Connection. The output resistance is defined by equation *3-21*:

$$r_o = r_{22} - \left(\frac{r_{12} r_{21}}{R_g + r_{11}} \right)$$

In terms of the internal transistor parameters, the output resistance becomes:

$$r_o = r_e + r_c - r_m - \left(\frac{(r_c - r_m) r_c}{(r_b + r_c + R_g)} \right) \qquad Eq.\ (4\text{-}16)$$

The variation of r_o with respect to the generator resistance R_g is illustrated for both transistors in Figs. 4-12 and 4-13. In the point-contact characteristic, the output resistance is positive over the range of R_g from 0 to approximately 50 ohms. When the generator resistance is increased beyond 50 ohms, r_o becomes negative, and gradually approaches a limiting value equal to r_{22} ($-11,850$ ohms) for large values of R_g. The output resistance of the junction transistor starts at a value of approximate-

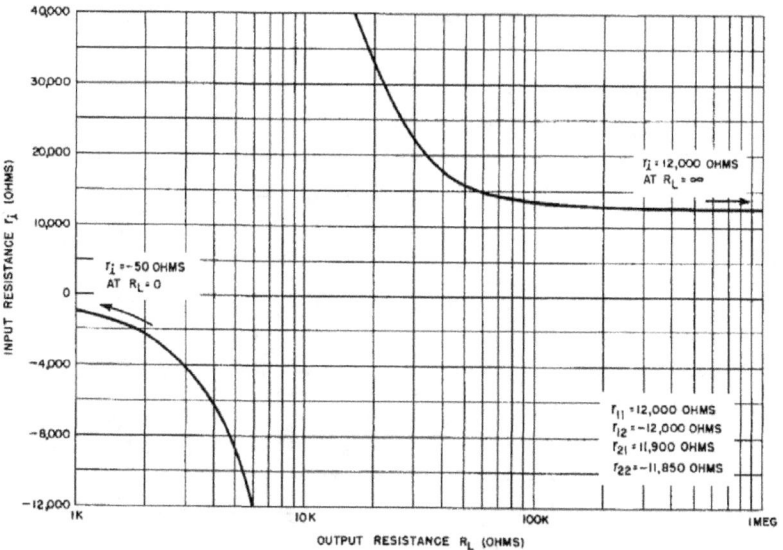

Fig. 4-10. Input resistance vs load resistance for typical point-contact transistor (grounded collector).

ly 75 ohms at $R_g = 0$ and gradually approaches a value equal to r_{22} (100,050 ohms) for large values of generator resistance.

As in the case of the grounded emitter, the grounded collector circuit using the point-contact transistor cannot be matched on an image

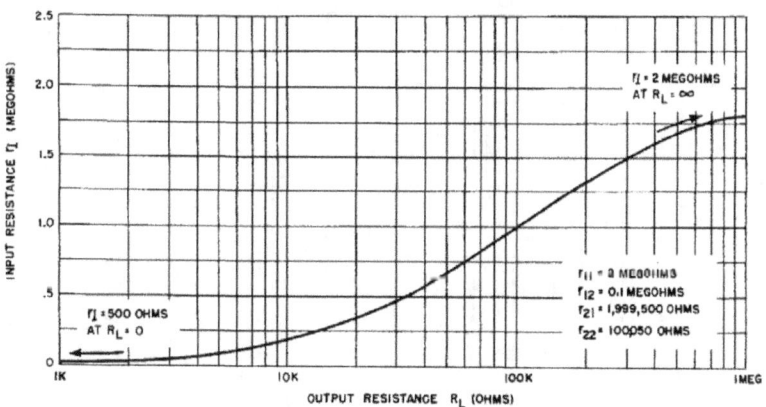

Fig. 4-11. Input resistance vs load resistance for typical junction transistor (grounded collector).

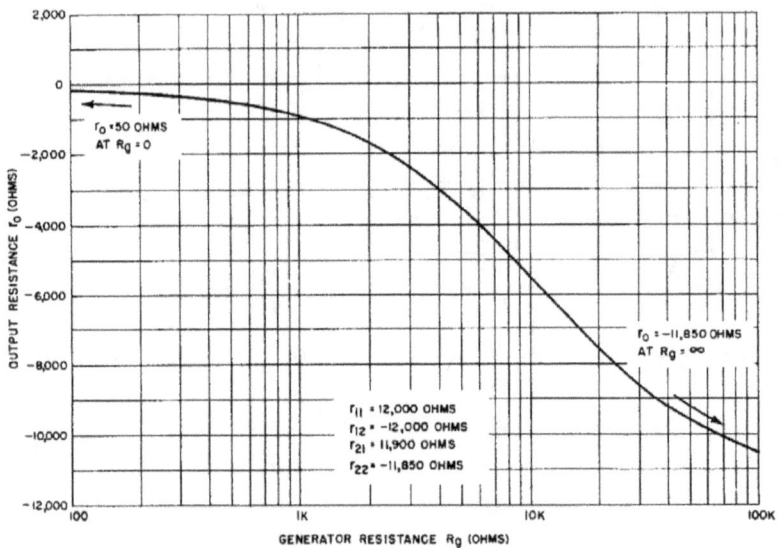

Fig. 4-12. Output resistance vs generator resistance for typical point-contact transistor (grounded collector).

basis without external modification, since the stability factor of this circuit is greater than one. However, the grounded collector does exhibit a unique characteristic when external resistance is added in the collector arm. For example, assume that a resistor R_c is added to the

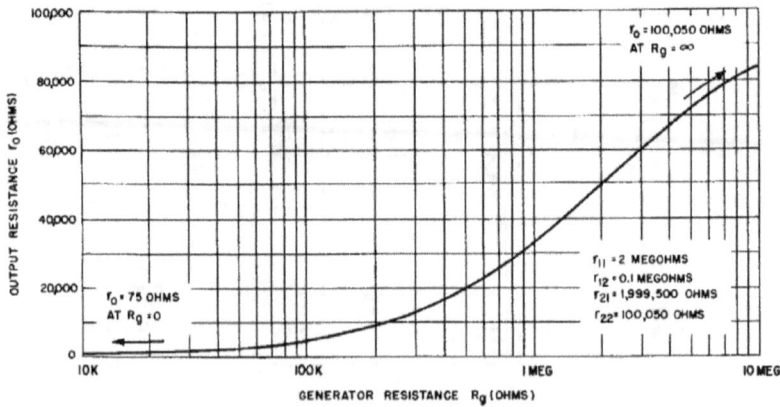

Fig. 4-13. Output resistance vs generator resistance for typical junction transistor.

collector arm so that $R_c + r_c = r_m$. For this modification,

$r_{11} = r_b + r_e + R_c$

$r_{12} = r_c + R_c - r_m = 0$

$r_{21} = r_c + R_c$ and

$r_{22} = r_e + r_c + R_c - r_m = r_e$

Since $r_{12} = 0$, the stability factor

$$\delta = \frac{r_{12}r_{21}}{r_{11}r_{22}} = 0$$

Thus, the modified circuit is stable. The input image matched resistance (equation 3-37) is then

$$r_1 = r_{11}\sqrt{1-\delta} = r_{11} = r_b + r_e + R_c$$

and the output image matched resistance (equation 3-41) becomes

$$r_2 = r_{22}\sqrt{1-\delta} = r_{22}$$

Notice also that $r_1 = r_1 = r_{11}$ and $r_o = r_2 = r_{22}$.

Thus, adding a suitable external resistor in the collector arm causes the circuit to act as a perfect buffer stage in which both the input and output resistances are independent of R_L and R_g.

Numerical values for the typical point-contact transistor modified to act as a buffer stage are:

$r_1 = r_1 = r_{11} = r_b + r_e + R_c = 100 + 11,900 + 12,000 = 24,000$ ohms;

$r_o = r_2 = r_{22} = r_e = 150$ ohms.

The image matched input and output equations can be applied to the junction transistor since its stability factor is always slightly less than one. A practical method to use in selecting values to be substituted in these equations indicates that r_1 should be chosen to equal 2 percent of r_{11}, and r_2 equal to 2 percent of r_{22}. The exact determination of the image matched resistances in the grounded collector circuit is not important, because the power gain is constant over a wide range of load resistances when the signal generator is matched to the input resistance.

In the junction transistor, numerical values for image matched resistances are

$$r_1 = r_{11}\sqrt{1-\delta} = 2,000,000 \sqrt{1 - \frac{100,000\,(1,999,500)}{100,050\,(2,000,000)}} = 33,200 \text{ ohms}$$

$$r_2 = r_{22}\sqrt{1-\delta} = 100,050 \sqrt{1 - \frac{100,000\,(1,999,500)}{100,050\,(2,000,000)}} = 1,670 \text{ ohms}$$

If the approximate values are used

$r_1 = .02r_{11} = .02\,(2,000,000) = 40,000$ ohms

$r_2 = .02r_{22} = .02\,(100,050) = 2,001$ ohms

Voltage Gain in the Grounded Collector Connection. The voltage gain, as defined in equation 3-24 by the general four-terminal parameters is:

$$VG = \frac{r_{21}R_L}{(R_L + r_{22})\,(R_g + r_{11}) - r_{12}r_{21}}$$

In terms of the internal transistor parameters for the grounded collector connection, the voltage gain becomes:

$$VG = \frac{r_e R_L}{(R_L + r_c + r_e - r_m)(R_g + r_e + r_b) - r_e(r_c - r_m)} \qquad Eq. \ (4\text{-}17) \ \bullet$$

Under the conditions for infinite input resistance and infinite current gain $(R_L = -r_{22})$, the voltage gain becomes:

$$VG = \frac{r_{21}(-r_{22})}{(-r_{22} + r_{22})(R_g + r_{11}) - r_{12}r_{21}} = \frac{r_{22}}{r_{12}} = \frac{r_c + r_e - r_m}{r_c - r_m}$$

For a perfect buffer stage $R_c + r_e - r_m = 0$. Thus, the voltage gain equation becomes

$$VG = \frac{(R_c + r_c) R_L}{(R_L + r_e)(R_g + r_b + r_e + R_c)}$$

The maximum voltage gain as defined by equation 3-25 is $VG = \dfrac{r_{11}}{r_{21}}$;

this becomes $\dfrac{r_c}{r_b + r_c}$. For the typical point-contact transistor, when $R_L = 15,000$ ohms and $R_g = 10$ ohms,

$$VG = \frac{r_{21} R_L}{(R_L + r_{22})(R_g + r_{11}) - r_{12}r_{21}}$$
$$= \frac{11,900 \, (15,000)}{(15,000 - 11,850)(10 + 12,000) - (-12,000)(11,900)} = .987$$

Under the conditions $R_L = -r_{22} = 11,850$ ohms

$$VG = \frac{r_{22}}{r_{12}} = \frac{-11,850}{-12,000} = .985$$

Under the conditions $R_c + r_c = r_m \, (R_c = 12,000$ ohms)

$$VG = \frac{(R_c + r_c) R_L}{(R_L + r_e)(R_g + r_b + r_e + R_c)}$$
$$= \frac{(12,000 + 11,900) \, 15,000}{(15,000 + 150)(10 + 100 + 11,900 + 12,000)} = .983$$

The maximum voltage gain $VG = \dfrac{r_{21}}{r_{11}} = \dfrac{11,900}{12,000} = .990$

For the typical junction transistor, $R_L = 100,000$ ohms, $R_g = 10$ ohms

$$VG = \frac{r_{21} R_L}{(R_L + r_{22})(R_g + r_{11}) - r_{12}r_{21}}$$
$$= \frac{1,999,500 \, (100,000)}{(100,000 + 100,050)(10 + 2,000,000) - 100,000 \, (1,999,500)} = .998$$

The maximum voltage gain $VG = \dfrac{r_{21}}{r_{11}} = \dfrac{1,999,500}{2,000,000} = .999$

Notice that in all of the above cases, the voltage gain is slightly less than unity. This is typical of the grounded collector connection.

Power Gain in the Grounded Collector Connection. The operating power gain as defined by equation 3-46 is:

$$G = \frac{4R_g R_L r^2_{21}}{[(R_g + r_{11})(R_L + r_{22}) - r_{12}r_{21}]^2}$$

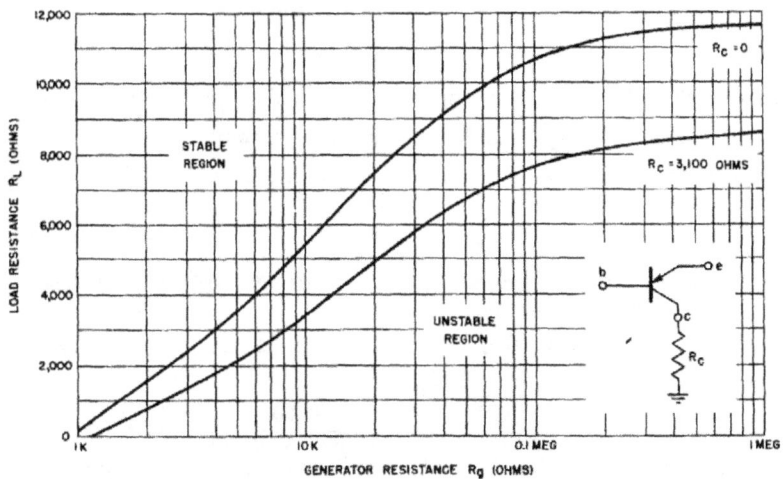

Fig. 4-14. Conditional stability characteristic (grounded collector).

As in the case of the grounded emitter connection, this gain equation is conditional for the point-contact transistor when the stability factor δ is greater than one. The conditional equation is:

$$(R_g + r_{11})(R_L + r_{22}) - r_{12}r_{21} > 0$$

Substituting the numerical values of the typical point-contact transistor into this equation:

$$(R_g + 12,000)(R_L - 11,850) + 12,000(11,900) > 0$$

The conditional stability characteristic is plotted in Fig. 4-14.

The grounded collector circuit can be stabilized by adding an external resistance R_e in the collector arm. For example, assume that a resistor $R_e = 3,100$ ohms is placed in series with r_c. The open-circuit parameters now become:

$$r_{11} = r_b + r_c + R_e = 100 + 11,900 + 3,100 = 15,100 \text{ ohms}$$
$$r_{12} = r_c + R_e - r_m = 11,900 + 3,100 - 23,900 = -8,900 \text{ ohms}$$
$$r_{21} = r_c + R_e = 11,900 + 3,100 = 15,000 \text{ ohms}$$
$$r_{22} = r_e + r_c + R_e - r_m = 150 + 11,900 + 3,100 - 23,900 = -8,750 \text{ ohms}$$

The conditional equation now becomes:

$$(R_g + 15,100)(R_L - 8,750) + 8,900(15,000) > 0.$$

This stabilized conditional stability line is also plotted on Fig. 4-14.

The maximum available gain, defined by equation 3-55:

$$MAG = \frac{r^2_{21}}{r_{11}r_{22}(1 + \sqrt{1 - \delta})^2}$$

can be applied to junction transistors, since the stability factor is not greater than one. In the grounded collector connection, since δ is always very near unity, this equation can be simplified as:

$$\text{MAG} = \frac{r_{21}{}^2}{r_{11}r_{22}\left(1 + \sqrt{1 - 1}\right)^2} = \frac{r_{21}{}^2}{r_{11}r_{22}} \qquad \textit{Eq. (4-18)} *$$

Notice that this result is nothing more than the product of the maximum voltage gain $\dfrac{r_{21}}{r_{11}}$ and the maximum current gain $\dfrac{r_{21}}{r_{22}}$. The maximum available gain in terms of the transistor internal parameters is:

$$\text{MAG} = \frac{r_c{}^2}{(r_b + r_c)\,(r_e + r_c - r_m)} \qquad \textit{Eq. (4-19)}$$

and since r_0 and r_e are negligible compared to the large values of r_c and $(r_c - r_m)$, the maximum available gain is:

$$\text{MAG} = \frac{r_c{}^2}{r_e\,(r_c - r_m)} = \frac{r_c}{r_c - r_m} \qquad \textit{Eq. (4-20)}$$

For the typical junction transistor

$$\text{MAG} = \frac{r_c}{r_c - r_m} = \frac{1,999,500}{1,999,500 - 1,899,500} = 20$$

Reverse Power Gain in the Grounded Collector Circuit. The grounded collector connection also has the unique ability to furnish power gain in the reverse direction. This characteristic might be anticipated on the basis of the equivalent circuit, since the internal generator $r_m i_e$ is common to both the input and output circuits, and the values of r_b and r_e are approximately equal. The equivalent circuit for the reverse connection is illustrated in Fig. 4-15. The resulting four-terminal parameters for this connection can be evaluated in terms of the internal

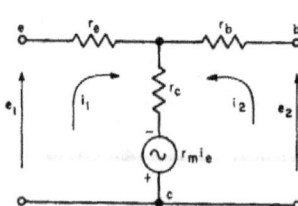

Fig. 4-15 (above). Equivalent "T" for reverse operation of grounded collector connection.

Fig. 4-16. (right). Transistor collector I_C-E_C characteristic, illustrating maximum limitations.

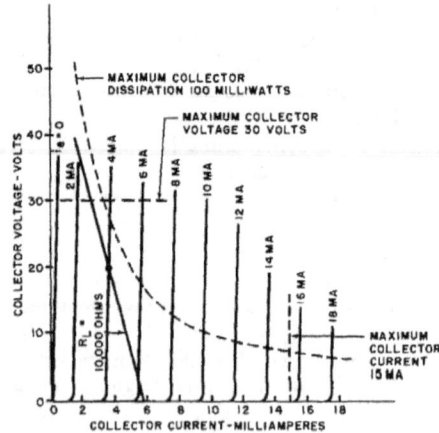

transistor parameters as before:

A. The input loop equation is:

$$e_1 + r_m i_e = i_1 (r_e + r_c) + i_2 r_c$$

Substituting $i_e = i_1$,

$$e_1 = i_1 (r_e + r_c - r_m) + i_2 r_c$$

when $i_2 = 0$, $e_1 = i_1 (r_e + r_c - r_m)$; then $r_{11} = r_e + r_c - r_m$, which is equal to r_{22} in the forward direction.

B. Using the same input loop equation, when $i_1 = 0$, $e_1 = i_2 r_c$, then

$$r_{12} = \frac{e_1}{i_2} = r_c,$$ which is equal to r_{21} in the forward direction.

C. The output loop equation is

$$e_2 + r_m i_e = i_1 r_c + i_2 (r_b + r_c);$$

Since $i_e = i_1$, $e_2 = i_1 (r_c - r_m) + i_2 (r_b + r_c)$

when $i_2 = 0$, $e_2 = i_1 (r_c - r_m)$; then $r_{21} = \dfrac{e_2}{i_1} = r_c - r_m$; which is equal to r_{12} in the forward direction.

D. Using the same output loop equation, when $i_1 = 0$, $e_2 = i_2 (r_b + r_c)$; then $r_{22} = e_2/i_1 = r_b + r_c$, which is equal to r_{11} in the forward direction.

Therefore, it can be seen that any of the equations derived for operation in the forward direction can be revised for use in the reverse direction by substituting r_{11} for r_{22}, r_{12} for r_{21}, r_{21} for r_{12}, and r_{22} for r_{11}. For example, the maximum available power gain in the forward direction, MAG $= \dfrac{r^2_{21}}{r_{12} r_{22}}$, becomes MAG $= \dfrac{r^2_{12}}{r_{22} r_{11}}$ in the reverse direction.

Comparison of Transistor Connections

The analyses of the three basic connections and their operating characteristics apply equally to both point-contact and junction transistors. However, due to the difference in comparative values of the internal transistor parameters, r_e, r_b, r_c, and r_m, the performance of the two basic transistor types is considerably different. In practice, the point-contact transistor is unstable, and has negative input and output resistances. On the other hand, the junction type is generally cheaper to produce, has better reliability, better reproducibility, higher available gain, and a lower noise figure than the point-contact type. It is safe to predict the gradual displacement of the point-contact transistor by the junction transistor in all but a few specialized applications, particularly since the frequency range of the junction type is steadily being increased by new manufacturing techniques. In view of this, the remainder of the book will deal primarily with the junction transistor, and unless specified, typical junction characteristics will be assumed.

At this point in the book all the basic design formulas have been derived for the three transistor connections. Thus, a comparison between the general characteristics of the three fundamental connections is now in order.

The grounded base connection is similar to the grounded grid circuit in electron tubes. This connection is characterized by low input resistance, high output resistance, and no phase inversion. Although its current gain is less than one, it provides respectable voltage and power gains. It is well suited for d-c coupling arrangements and for preamplifiers that require a low input and high output impedance match.

The grounded emitter circuit is the transistor equivalent of the grounded cathode connection in the vacuum tube circuit. This transistor connection is the most flexible and most efficient of the three basic connections. The grounded emitter connection reverses the phase of the input signal. Its matched input resistance is somewhat higher than that of the grounded base connection; its matched output resistance is considerably lower. The grounded emitter usually provides maximum voltage and power gain for a given transistor type.

The third connection, the grounded collector, is the transistor equivalent of the grounded-plate vacuum tube. It is characterized by a voltage gain that is always slightly less than unity. Its current gain is in the same order as that of the grounded emitter. It has a relatively low output resistance, a high input resistance, and does not produce phase reversal. It offers low power gain, but is capable of supplying reverse power gain. The grounded collector circuit is used primarily as a matching or buffer stage.

Transistor Limitations

Maximum Limits. To use the transistor in practical circuits, it is necessary to be aware of its limitations. First, the transistor has limited power-handling capabilities. (The maximum power dissipation rating of a transistor is always specified in the manufacturer's rating sheet.) Because the dissipation rating is relatively low, the operating temperature of the transistor is usually kept in the general temperature range of 50°C to 60°C. Relatively low ambient temperatures are also desirable because germanium is temperature sensitive, and behaves erratically at higher temperatures. In addition, the operating range is limited by the maximum allowable collector voltage (a function of the Zener voltage, previously discussed), and the maximum collector current. (The values of these latter factors are also specified in the manufacturer's rating sheets.)

Figure 4-16 illustrates the three maximum limitations of a typical transistor having the following specified ratings: maximum collector dissipation — 100 milliwatts; maximum collector voltage — 30 volts; and maximum collector current — 15 milliamperes. The useful region of the collector current-voltage characteristics is necessarily limited to the area contained within these boundaries. In circuit application, none of the limiting factors can be ignored; exceeding any of the limits may damage the transistor. For example, assume that the transistor illustrated in Fig. 4-16 is operated as follows: collector current $I_c = 4$ milliamperes, collector voltage $E_c = 20$ volts, load resistance $R_L = 10,000$ ohms. Assume also that the a-c input signal causes a collector current variation of ± 2 milliamperes. Thus, the output signal varies along the load line between the collector current limits of 2 to 6 milliamperes. The collector current never exceeds the maximum limit of 15 milliamperes, and at peak signal the collector dissipation is 40 volts times 2 milliamperes (80 milliwatts), which is well within the maximum power limits. However, the collector voltage is now 10 volts greater than the allowable limit of 30 volts. The transistor, therefore, would not be suitable for the assumed operation.

Minimum Limits. The minimum limits of the transistor are generally not critical in practical cases. The minimum collector voltage is set by the non-linear portion of the characteristic curve, which is not reached until the collector voltage is reduced to a few tenths of a volt. The minimum collector current must be greater than the saturation current I_{co}, which is considerably less than 100 microamperes in most junction types. The error introduced by assuming the minimum limits to be $E_c = 0$ and $I_c = 0$ is generally negligible.

Transistor Noise. The minimum signal that can be applied to a transistor is limited by the internal noise generated by the transistor. Since the transistor does not require cathode heating (one of the major noise sources in the vacuum tubes), it is inherently capable of operating at lower noise levels than its vacuum tube brother. At present, the junction transistor is equal to the vacuum tube, insofar as its noise characteristics are concerned. The noise level of the point-contact types is between 15 and 30 db higher.

There is some confusion in the field as to what is meant by the manufacturers' specifications on noise limits. This confusion is caused by the various manners in which the noise level is specified. The noise level, when specified "with reference to thermal noise," tells the most about the transistor, because the reference value is reasonably fixed. The noise factor on this thermal basis is the ratio of the noise power delivered to a load compared to the power delivered if the only source of noise were the thermal noise of the signal generator. A second method of noise specification is the "signal-to-noise ratio." The noise figure on this basis does not tell as much about the transistor as the first method, because the signal is not at a constant level. Another method is specification of noise in db above one milliwatt (dbm). This method is least useful since it neither specifies amplifier gain nor bandwidth.

The noise figure of the junction transistor is about 10 db above thermal noise at 1,000 cps; by selection, values as low as 5 db have

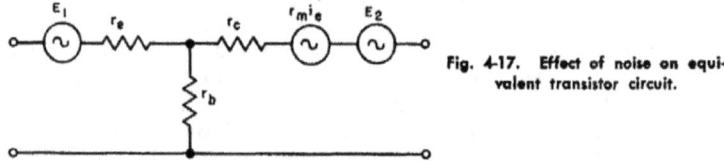

Fig. 4-17. Effect of noise on equivalent transistor circuit.

been found. These noise levels are comparable with those of the best vacuum tubes available. In general, the noise energy in the transistor is concentrated in the lower frequencies and, as might be expected, the noise factor decreases as the operating frequency is increased. The noise factor is affected by the operating point and the signal generator resistance. It appears to be lowest both at low values of collector voltage and when the generator resistance R_g is equal to the input resistance r_1. In general, transistors with large collector resistance have a low noise level. Figure 4-17 illustrates the equivalent circuit of the grounded base connection, and includes the equivalent voltages E_1 and E_2 introduced by transistor noise.

Testing Transistors

Basic Circuits. Although the manufacturer's data sheets for transistors are very useful in preliminary paper studies of circuits, it is often necessary to make direct transistor measurements. The block diagram of Fig. 3-8 illustrates the basic circuits for measuring the a-c open-circuit parameters r_{11}, r_{12}, r_{21}, and r_{22}. (Methods for measuring a and I_{co} are indicated later in this section.) The following general rules aid the experimenter in obtaining reasonably accurate results for all measurements.

1. Use an accurate meter calibrated for the appropriate operating range. This is required since the transistor operates on comparatively small values of current and voltage.

2. Measure the d-c bias voltages with a very high resistance voltmeter, to avoid meter-shunting effects. Shunting errors are particularly noticeable in the collector circuit which may have resistance of several megohms.

3. Connect the test signal (usually 1,000 cps) through a step-down transformer that has an impedance ratio in the order of 500:1. This keeps the measurements independent of R_g and, at the same time, permits a low signal input without requiring a low oscillator gain control setting.

4. Measure all calibrating resistors with an accurate bridge, or use a calibrated resistor decade box for the resistors.

5. Check the waveform with an oscilloscope. The waveform quickly indicates reversed bias connections and overloads.

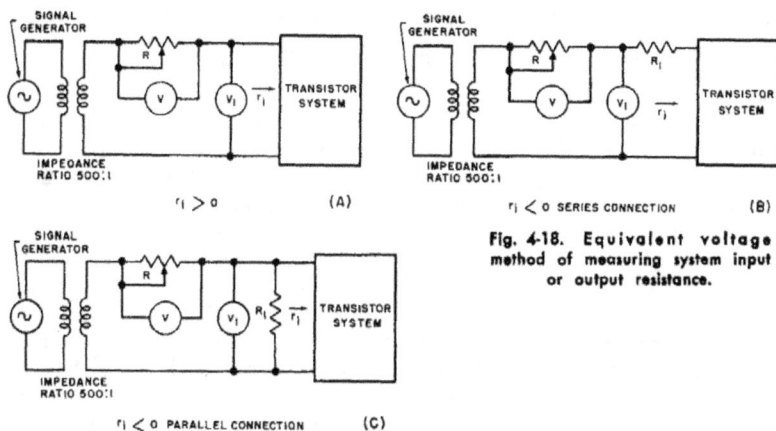

Fig. 4-18. Equivalent voltage method of measuring system input or output resistance.

Equal Voltage Method. The equal voltage method is a quick way of determining the input or output resistance of a system when the equipment is limited. This connection is illustrated in Fig. 4-18 (A).

Resistor R is a calibrated decade box or a helipot in series with the effective input resistance of the system under test. Resistor R is adjusted until its voltage drop V is equal to the input voltage V_1. Since the arrangement is a simple series circuit, the input resistance r_1 is then equal to R.

Figure 4-18 (B) illustrates the equal voltage method for measuring a negative resistance. In this case, a calibrated resistor R_1 having a larger absolute value than that of the negative resistance is connected in series with r_1. Again resistor R is adjusted until $V = V_1$, for which $R - R_1 = r_1$. For example, suppose a point-contact transistor is operating in its negative resistance region. When a resistor $R_1 = 2,000$ ohms is placed in series with the input, it brings the circuit into its positive input region (stable operation). When the connection of Fig. 4-18 (B) is set up, $R = 1,225$ causes V to equal V_1. Then $r_1 = R - R_1 = 1,225 - 2,000 = -775$ ohms.

Notice that this latter arrangement requires that R be greater than the absolute value of r_1. If the only calibrated resistors available are low in value, the parallel method illustrated in Fig. 4-18 (C) can be used. The procedure is the same as before except that when $V = V_1$, R is equal to R_1 and r_1 in parallel,

$$R = \frac{R_1 r_1}{R_1 + r_1}$$

which in terms of the input resistance becomes:

$$r_1 = \frac{R R_1}{R_1 - R}$$

For the same transistor measured above, if $R_1 = 500$ ohms, R is adjusted to 1,408 ohms, at which time $V = V_1$. Then $r_i = \dfrac{(1408)\,(500)}{500 - 1408} =$ -775 ohms.

Transistor Test Sets. Several elaborate transistor test sets are available commercially. These testers are useful for large-scale experimental work, since they incorporate means for completely evaluating the characteristics of all types of point-contact and junction transistors, and do not require external test equipment and meters. The home experimenter and the lab technician, however, can get satisfactory results on breadboards, based on the techniques described on the previous pages.

In checking transistors during maintenance and repair, it is not necessary to check all the transistor parameters. A check of two or three of the performance characteristics will determine quickly whether a transistor needs to be replaced.

Figure 4-19 illustrates a transistor check circuit which will measure the current gain and saturation current with reasonable accuracy. The operation procedure and general functional description of the circuits follows:

1. With switch SW2 in the calibrate (CAL) position and switch SW1 in the current gain (a) position, adjust the signal gain of the

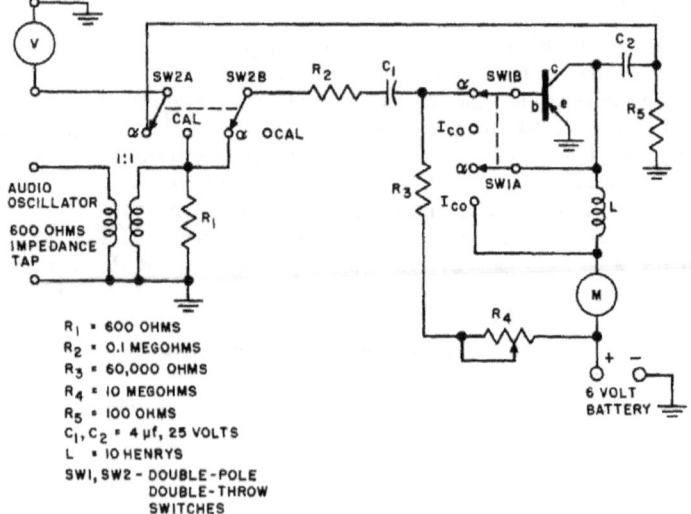

R_1 = 600 OHMS
R_2 = 0.1 MEGOHMS
R_3 = 60,000 OHMS
R_4 = 10 MEGOHMS
R_5 = 100 OHMS
C_1, C_2 = 4 μf, 25 VOLTS
L = 10 HENRYS
SW1, SW2 - DOUBLE-POLE DOUBLE-THROW SWITCHES

Fig. 4-19. Transistor tester for measuring a and I_{co}.

audio oscillator for one volt across resistor R_1. Now throw SW2 to the current gain (a) position. The signal is now connected to the base of the transistor through resistor R_2 and the d-c blocking capacitor C_1. Since resistor R_2 is 100,000 ohms, the base and emitter resistances of the transistor are negligible; a-c base current i_b, 10 microamperes.

2. The d-c base current bias is controlled by resistors R_3 and R_4, which permit a variation of from about 1 to 100 microamperes. R_4 is adjusted until the collector d-c bias current, measured by meter M, is one milliampere.

3. Practically all of the a-c output appears across the 100 ohm resistor R_5, because of the high impedance of choke coil L (over 60,000 ohms at 1,000 cps), and the high output resistance of the transistor (usually more than a megohm). The output voltage across R_5 is $a\,i_b R_5$, and since $i_b = 10$ microamperes, $R_5 = 100$ ohms, this voltage equals $.001a$. The value of a may vary from 10 to 100. The a-c voltage may, therefore, range from .01 to .1 volt. Thus, the current amplification can be taken directly on a low scale of a good voltmeter.

Due to the comparatively low value of R_5, the measured reading closely approximates the maximum current gain $a = r_{12}/r_{21}$. This value of current gain for the grounded emitter connection can be converted into approximately equivalent values for the grounded base and grounded collector circuits by means of the following conversion formulas:

$$a_{GB} = \frac{a_{GE}}{1 + a_{GE}} \qquad\qquad Eq.\ (4\text{-}21)$$

and

$$a_{GC} = \frac{a_{GE}}{a_{GB}} \qquad\qquad Eq.\ (4\text{-}22)$$

where a_{GE} = maximum current gain for grounded emitter connection; a_{GB} = maximum current gain for the grounded base connection; and a_{GC} = maximum current gain for the grounded collector connection. These relationships are derived by neglecting r_e and r_b in comparison with r_m, r_c and $(r_c - r_m)$. Error in this approximation is negligible.

For example, assume that a transistor is tested in the circuit of Fig. 4-19 and produces a reading of .022 volt on the a-c output voltmeter connected across R_5. The current gain

$$a_{GE} = \frac{.022}{.001} = 22;\ a_{GB} = \frac{22}{1 + 22} = .96;\ a_{GC} = \frac{22}{.96} = 22.9$$

The saturation current is read directly on the milliammeter M if switch SW1 is now placed in the I_{co} position. This switch opens the base lead, removing the bias, and also shorts out the inductor L so that the six-volt battery is across the emitter and collector electrodes.

The circuit as shown is only suitable for N-P-N junction transistors, but can be modified easily for the P-N-P type by incorporating a switch to reverse the battery, the meter connections, and the d-c blocking electrolytic capacitors.

Chapter 5
TRANSISTOR AMPLIFIERS

This chapter deals with the design and operation of the transistor as a low-frequency amplifying device based on the transistor characteristics and limitations discussed in the preceding chapters. Since it is impracticable to cover every useful type of connection, the emphasis in this section is on fundamental illustrations, such as choosing the transistor d-c operating point, stabilizing methods, matching, direct coupling, and cascading class A and B single-ended and push-pull transistor amplifiers. Some of the unique properties of transistors that are attained by the symmetrical operation of the N-P-N and P-N-P types in the same circuit are also considered.

Grounding the Transistor System

Some confusion exists about which electrode should be connected to ground in a transistor system. The basic reason for the difficulty lies in the terminology: grounded base, grounded emitter, and grounded collector. Actually, these designations do not refer to the circuit ground, but only specify which of the three electrodes is common to both the input and output circuits. A better way to specify the three basic connections would be: common base, common emitter, and common collector. These latter designations are used by many authorities. In general, the system ground can be made at any convenient point in the circuit, without consideration to the type of connection.

The D-C Operating Point

Limitations, Supply Voltage and Load. As in the case of the vacuum tube, the problem of designing a transistor amplifier is somewhat

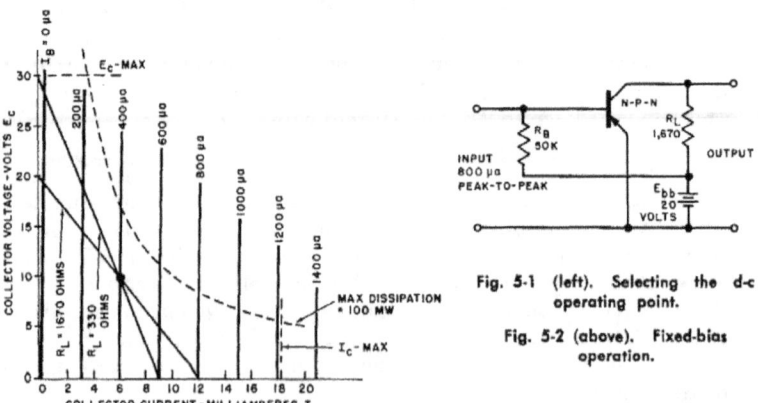

Fig. 5-1 (left). Selecting the d-c operating point.

Fig. 5-2 (above). Fixed-bias operation.

simplified if the a-c signal is treated independently of the d-c operating point. The first step in the design could logically be the selection of the d-c operating point. (Actually, three separate conditions must be fixed; the operating point, the load line, and the supply voltage. In general, the selection of any two automatically limits the determination of the third.) The d-c operating point may be placed anywhere in the transistor characteristics, limited however by the collector maximums of voltage, current, and power dissipation. The final selection of the operating point is based primarily on the magnitude of the signals to be handled.

Suppose, for example, a transistor, whose characteristics are illustrated in Fig. 5-1, is to be used with its operating point set at $E_c = 10$ volts, $I_c = 6$ ma. Assume, also, that the maximum limits of the transistor are $I_c = 18$ ma, $E_c = 30$ volts, and collector dissipation = 100 milliwatts, as shown enclosed by the dotted lines. The supply voltage required is the value at the intersection of the load line and the collector voltage axis. Thus, for a fixed load of 1,670 ohms, the necessary supply voltage is 20 volts. If, however, the supply voltage is fixed, then the load resistance is determined by the line joining both the supply voltage (E_c at $I_c = 0$) and the operating point. As an illustration, assume the supply voltage is to be E_{bb} fixed at 30 volts. The resulting load resistance

$$R_L = \frac{E_{bb} - E_c}{I_c} = \frac{30 - 10}{6 \times 10^{-3}} = 3,330 \text{ ohms.}$$

For any selected operating point there are many combinations of load resistance and supply voltage that will permit the load line to pass through the d-c operating point.

The usual problem is one in which both the load and supply voltages are fixed. The problem then resolves itself into a choice of the operating point. In Fig. 5-1, for the conditions $R_L = 1,670$ ohms, and $E_{bb} = 20$ volts, the d-c operating point may be placed anywhere along the load line. It is usually desirable to design the amplifier for maximum signal handling capacity. In this case, then, the d-c operating point should be midway between the extreme limits of the base current, namely 0 and 800 microamperes. The choice of $I_b = 400$ microamperes sets the operating point for maximum signal capacity at $I_c = 6$ma, and $E_c = 10$ volts.

Fixed Bias. The collector bias conditions, then, fix the d-c bias current I_b of the input base electrode; conversely, the base bias current fixes the collector bias for a given load and supply voltage. The desired base bias current can be obtained by connecting a resistor between the base and the collector terminal of the supply voltage as shown in Fig. 5-2. For $E_{bb} = 20$ volts, and $I_b = 400$ microamperes, the total series resistance is $\dfrac{E_{bb}}{I_b} = \dfrac{20}{400 \times 10^{-6}} = 50,000$ ohms.

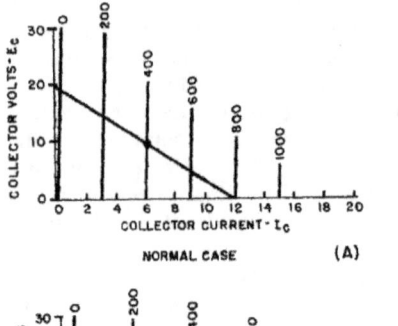

NORMAL CASE (A)

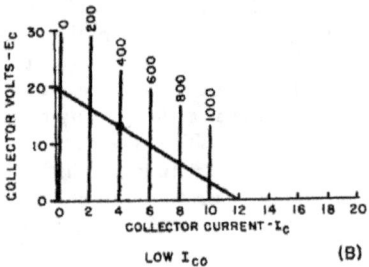

LOW I_{co} (B)

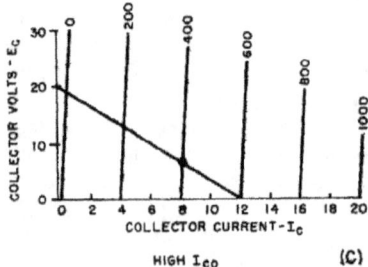

HIGH I_{co} (C)

Fig. 5-3. Variation of operating point.

This value includes the emitter to base resistance, but since $r_e + r_b$ is generally only a few hundred ohms, they can be neglected. The resulting circuit, with the calculated values, is illustrated in Fig. 5-2 for a N-P-N transistor. If the same characteristics were applied to a P-N-P type, the only circuit change would be a reversal of the supply battery potentials. The transistor bias indicated in this figure is called *fixed bias*.

Self-Bias. Unfortunately, transistors are temperature sensitive devices; in addition some variation usually exists in the characteristics of transistors of a given type. These factors may cause a displacement of the constant base current lines along the collector current axis. Figure 5-3 illustrates the effect of this variation; the abnormal cases are purposely exaggerated. Notice the effect of this shift on the relative positions of the d-c operating point. In the low I_{co} unit (Fig. 5-3B) the collector voltage is too high; in the high I_{co} unit (Fig. 5-3C) the collector voltage is too low. To overcome this, the circuit needs degeneration, similar to that produced by an unbypassed cathode bias resistor in a vacuum-tube circuit. In transistor circuitry, this method of degeneration is a form of automatic control of the base bias, known as *self bias*.

A simple method for establishing automatic control of the base bias requires the base bias resistor to be tied directly to the collector, as in Fig. 5-4. Thus, if the collector voltage is high (Fig. 5-3B), the base current is increased, moving the d-c operating point downward

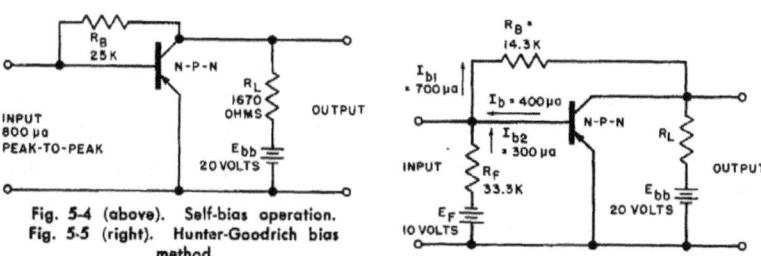

Fig. 5-4 (above). Self-bias operation.
Fig. 5-5 (right). Hunter-Goodrich bias method.

along the load line; conversely, if the collector voltage is low (Fig. 5-3C), the base bias current is decreased, moving the d-c operating point upward along the load line. The value of the selected base bias resistor is different in the self-bias case from that computed in the fixed-bias connection. For self bias, the resistor is tied to the collector voltage, which in this case is 10 volts. Then $R_B = \dfrac{E_c}{I_b} = \dfrac{10}{400 \times 10^{-6}} = 25,000$ ohms. The base bias resistor performs the double duty of determining the value of I_b and preventing those excessive shifts in the collector d-c operating point due to temperature change and transistor interchange. The principal limitation of self bias is that it still allows some variation of the d-c operating point, since the base bias resistor is fixed by the required operating point, and the stabilization produced by it is only a secondary effect. In addition, self bias also introduces a-c negative feedback which reduces the effective gain of the amplifier. Despite its limitations, however, self bias is very useful and works well in many applications.

The importance of temperature stability with respect to the d-c operating point cannot be taken lightly. One of the effects of a temperature rise is to increase the saturation current I_{co}, which, in turn, increases the collector dissipation. The increased collector dissipation increases the temperature, which increases I_{co}, and so on. Thus, poor temperature stability almost certainly will cause transistor burnouts, particularly if the transistor is operated near its maximum dissipation limit.

Hunter-Goodrich Bias Method. A method of establishing tighter control on the base bias current illustrated in Fig. 5-5 is the Hunter-Goodrich method. This involves the addition of a fixed base bias operating in the reverse direction of the normal self bias. The fixed bias is introduced by resistor R_F and separate voltage supply E_F. To overcome this reversed fixed bias, the self bias resistor R_B must be decreased to maintain the same base bias current. The reduced value of R_B increases the available negative d-c feedback from the collector circuit, thus providing greater transistor stability.

As in the preceding cases, the effect of the base and emitter circuit resistances $(r_e + r_b)$ can be neglected in the calculations. The values of R_F and E_F depend upon the value of fixed bias desired. For example, assume that a fixed bias value I_{b2} of 300 μa will provide the additional stability needed, and a battery $E_F = 10$ volts is available. Then $R_F = \dfrac{E_F}{I_{b2}} = \dfrac{10}{300 \times 10^{-6}} = 33,300$ ohms. The current through the self bias resistor R_B is $I_{b1} = I_b + I_{b2} = 400 + 300 = 700$ μa; then $R_B = \dfrac{E_F}{I_{b1}} = \dfrac{10}{700 \times 10^{-6}} = 14,300$ ohms.

In comparison, $R_B = 25,000$ ohms in the simple self bias case. Since the input resistance of the transistor is small compared to R_F, practically all of the stabilizing current flows into the base-emitter circuit.

The Hunter-Goodrich bias method is extremely useful when a high degree of circuit stability is needed. Its particular disadvantage is that it requires two separate battery supplies.

Self Bias Plus Fixed Bias. One method of obtaining additional stabilization with only one battery is shown in Fig. 5-6 (A), the basic features of which are often used in transistor power stages. The fundamental differences between this circuit and the preceding fixed plus self bias method are the interchange of R_L and E_{bb}, and the connection of the reverse bias resistor R_F into the collector circuit. Interchanging the supply battery and the load resistor provides two points at which variations in collector voltage will appear. However, this interchange does not affect the d-c operation of the circuit. Connecting R_F, as illustrated, produces essentially the same result as the Hunter-Goodrich arrangement, except that the reverse bias is no longer fixed. If the previous circuit constants are desired: $R_F = \dfrac{E_c}{I_{b2}} = \dfrac{10}{300 \times 10^{-6}} = 33,300$ ohms. All the other values remain the same.

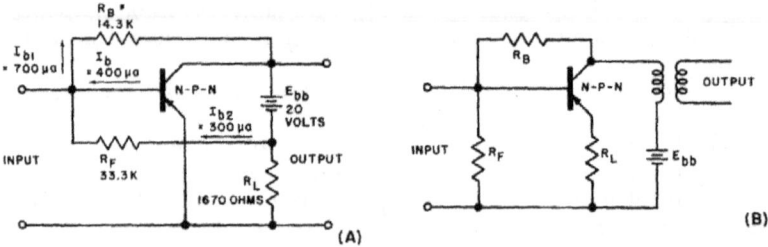

Fig. 5-6. (A) Stabilization of d-c operating point with one battery. (B) Typical power output stage.

In power amplifier circuits, the load usually consists of a transformer plus an additional stabilizing resistor. Figure 5-6 (B) illustrates one possible form of this arrangement for use as a transistor power amplifier stage.

A disadvantage of this bias method is that the d-c degeneration feedback is reduced, due to the shunting effect of resistor R_F, thus reducing the stabilization. On the other hand, this method provides for greater stability than does the simple self-bias method. It provides less stability than the Hunter-Goodrich method, but requires only one battery supply.

Current Sources. Notice that all the bias requirements are supplied by conventional batteries, which act as constant voltage sources. At this point the conscientious reader may wonder if this does not conflict with the statements in earlier chapters that transistors are current-operated devices. Actually, the term "current source" is more than just a mathematical concept. The practical aspect can be shown as follows: Assume that a six-volt battery with negligible internal resistance is connected to a variable load resistance. Except for very low values of load, the battery terminal voltage remains constant as long as the battery remains fully charged. Now assume that a one megohm resistor is connected in series with the battery and the load resistor. In this case the current remains reasonably constant while the load resistance is varied from zero to about 0.1 megohm. Thus, the addition of a series resistor has converted the constant voltage supply into a constant current source over a fairly wide range of load resistance values. The range depends upon the value of the series resistor. Figure 5-7 illustrates the basic equivalent interchanges of supply sources. Mathematically, all that is involved is

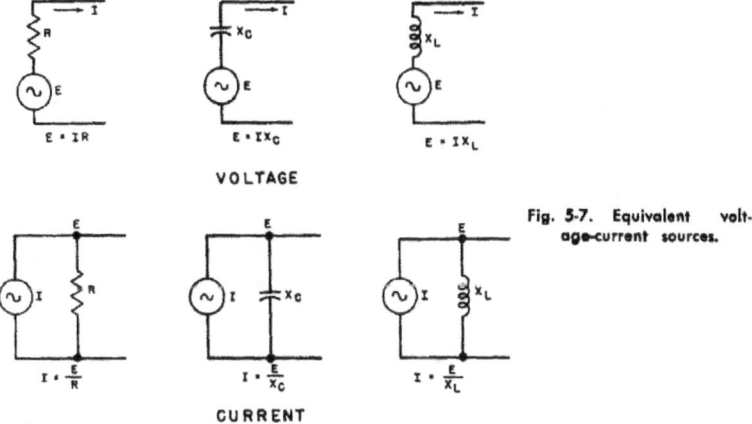

VOLTAGE

CURRENT

Fig. 5-7. Equivalent voltage-current sources.

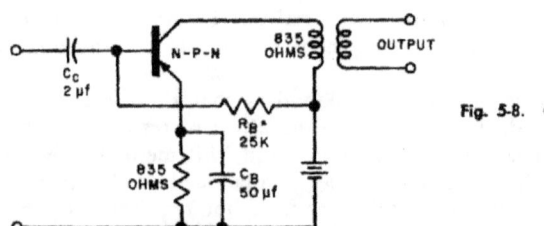

Fig. 5-8. Class A amplifier.

the movement of the impedance proportionality constant from one side of the equation to the other.

That these circuits are equivalent can be shown by a simple example. Take the case of a six-volt battery in series with a resistor $R = 1$ megohm and a load $R_L = 1$ megohm. Then the load current equals $\dfrac{E}{R + R_L} = \dfrac{6}{(1+1) \times 10^6} = 3$ microamperes. The equivalent circuit on a current basis is a current generator $I = \dfrac{E}{R} = \dfrac{6}{1 \times 10^6} = 6$ microamperes, which is shunted by both a resistor $R = 1$ megohm, and

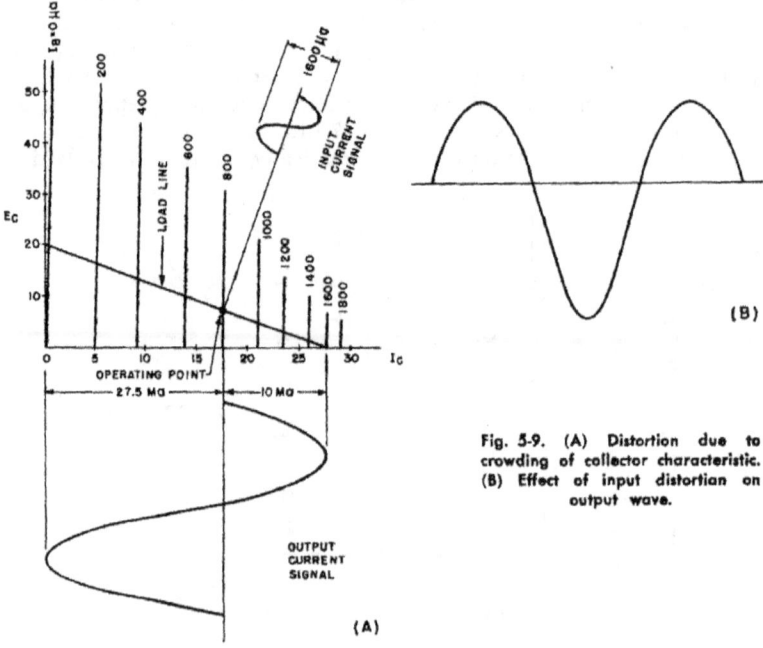

(A)

(B)

Fig. 5-9. (A) Distortion due to crowding of collector characteristic. (B) Effect of input distortion on output wave.

a load $R_L = 1$ megohm. Now the load current equals the source current less the amount shunted by resistor R. Since R and R_L are in parallel, the voltage drop across each resistor must be the same, and the load current equals

$$\frac{IR}{R + R_L} = \frac{6 \times 10^{-6} (1 \times 10^6)}{(1 + 1) \times 10^6} = 3 \text{ microamperes.}$$

This checks with the previous result. The same procedure can be used to convert an a-c voltage source into an a-c current source.

Class A Amplifiers

Basic Circuitry — Efficiency Stabilization. Figure 5-8 represents a typical Class A transistor amplifier using d-c operating biases as described in the preceding paragraphs. The stabilizing resistor in the emitter circuit is made equal to the load impedance in this case. This condition provides maximum protection against variations in I_{co}, since the power available from the battery is effectively limited to the maximum collector dissipation. While the arrangement shown in Fig. 5-8 prevents transistor damage due to excessive collector variations, half of the d-c power is dissipated across the stabilizing resistor. The maximum efficiency of a class A transistor is 50%; using a stabilizing resistor whose value is equal to the load resistor reduces the efficiency to a maximum of 25%.

In general, this amount of stability control is needed only in mass production applications if transistors having a wide tolerance range are to be used. Actually, the reproducibility of transistor characteristics has improved rapidly during the past few years. There is no reason why this trend should not continue, and eventually permit the attainment of amplifier efficiency values very close to the theoretical maximum. In most circuits, between 5 and 10% of the collector d-c power $(E_c I_c)$ is satisfactory for normal stabilization. In the circuit shown in Fig. 5-8, for example, a resistance of 100 ohms between the emitter and ground would be sufficient.

Bypass and Coupling Capacitors. If the stabilizing resistor in the emitter lead is unbypassed, the amplifier gain is decreased. This is similar to the action of an unbypassed cathode resistor in a vacuum-tube amplifier. A value of 50 μf works out well for the bypass capacitor in most audio frequency applications. The self bias resistances from collector to base may also be suitably bypassed to avoid a-c degeneration. In cascaded stages, the load is unusually low, and the a-c collector voltage is also low. In this case, the bypass capacitor can be omitted with only a slight loss in the stage gain.

The value of the coupling capacitor C_c must be large enough to pass the lowest frequency to be amplified. Usually a maximum drop of 3 db in gain is permitted. At this value, the reactance of the coupling capacitor is equal to the input resistance of the stage. Since the input resistance is low for the grounded emitter and grounded base connections, relatively high capacity coupling condensers are required. For example: What is the minimum value of C_c necessary for coupling into a stage whose input resistance $r_1 = 500$ ohms, if a frequency response down to 100 cps is required? At 100 cps,

$$X_c = \frac{1}{2\pi f c_c} = r_1 = 500 \text{ ohms, and } C_c = \frac{1}{2\pi f X_c} = \frac{1}{2\pi (100) (500)} = 3.2 \; \mu f.$$

In typical circuits, the required value of the coupling capacitor varies from 1 to 10 μf.

Distortion. Another characteristic of a transistor amplifier that should be mentioned is the harmonic distortion. If the circuit is wired, the distortion can be measured directly, using a suitable wave analyzer or distortion meter. In addition, distortion can be calculated under given operating conditions from the collector characteristics, using the same methods as in vacuum-tube amplifiers. These methods are described in detail in most radio engineering handbooks. The total harmonic distortion is about 5% in the typical transistor amplifier. It is caused mainly by the decreased spacing between the collector current-collector voltage curves for equal changes in base current. This crowding effect occurs at the higher values of collector current. Figure 5-9 (A) is an exaggerated illustration of this type of distortion in transistor circuits.

If the input resistance of the amplifier is high compared to the source impedance, another type of distortion, due to variations in the input circuit, is introduced. In the region of low collector current, the input resistance increases, thus reducing the amplitude of the input signal. In the high region of the collector a-c current cycle, the input resistance decreases, thus increasing the amplitude of the input signal. This type of non-linear distortion is illustrated in Fig. 5-9 (B).

Since the two major types of distortion described above have opposite effects, it will be possible to counteract one with the other by adjusting the value of the signal generator impedance. In troubleshooting multi-stage transistor amplifiers, an output waveform similar to that indicated in Fig. 5-9 (B) would probably indicate a defect in one of the preceding stages.

When computing the harmonic distortion of a transistor amplified by conventional vacuum-tube graphical methods, the computed value is generally in the order of 1% less than the measured value. This is caused by the assumption that the signal generator resistance is negligible, a condition seldom realized in low input resistance transistor circuits. The source resistance in vacuum-tube amplifiers does not affect the determination of the harmonic distortion, since the grid current is zero. However, the equivalent parameter in transistor amplifiers, the base-emitter voltage, is not zero. The effect of the source may be taken

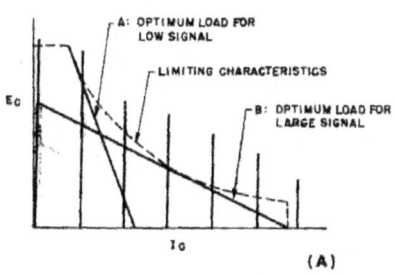

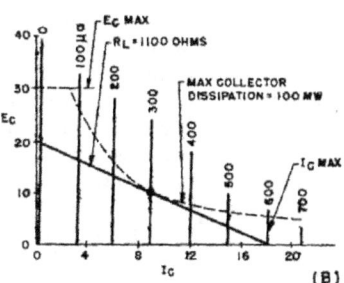

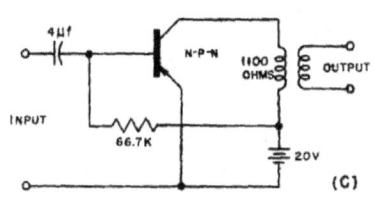

Fig. 5-10. (A) Load lines for maximum power. (B) Determination of optimum load. (C) Power amplifier circuit.

into account by considering it as part of the base resistance. However, in most applications, it is more than satisfactory to simply add 1% to the calculated value of percentage distortion.

Maximum Output Conditions. Since the power handling capacity of the transistor is small compared to that of the vacuum tube, it is usually necessary to drive the transistor to its maximum limits. When a transistor amplifier is designed for maximum power, as in Fig. 5-10 (C), it is properly termed a power amplifier, although the actual power involved may only amount to a hundred milliwatts. To obtain maximum power, the load line is selected to include the maximum possible area concurrent with the fixed limitations of maximum collector dissipation, current, and voltage. The ideal load for maximum power would be one which followed exactly the transistor limit boundaries illustrated in Fig. 5-1.

In the practical case, it is usually necessary to settle for a load line that is tangent to the limiting characteristic line. Since this curve is non-linear, there are several possible choices of load. The final choice depends primarily on the signal requirements of the circuit. Figure 5-10 (A) shows the two extreme cases: line A is the optimum load for a small signal input; line B is the optimum load for a large signal input. Since, however, the supply voltage is usually specified, the load line chosen is the one that is tangent to the limiting curve and that passes through the specified supply voltage (E_c at $I_c = 0$).

The calculations for determining the conditions for maximum output in a power amplifier stage, assuming a transistor with the characteristics illustrated in Fig. 5-10 (B), are as follows: Assume a battery supply $E_{bb} = 20$ volts. This determines a point on the load line for $I_c = 0$. Now hold one end of a straight edge on this point and swing the edge until it just touches a point on the maximum collector dissipation line. Draw a line through the two points. This is the optimum load line for the given conditions. The maximum input signal, the d-c bias resistors, and the distortion can all be computed directly from the figure by means of the methods discussed in the preceding paragraphs:

$$R_L = \frac{E_c \,(\text{at } I_c = 0)}{I_c \,(\text{at } E_c = 0)} = \frac{20}{.0182} = 1,100 \text{ ohms.}$$

$$R_B = \frac{E_{bb}}{I_b} = \frac{20}{300 \times 10^{-6}} = 66,670 \text{ ohms.}$$

Using the standard equation derived for vacuum-tube circuits, the transistor a-c output P_{ac} is one-eighth of the product of the peak-to-peak collector voltage and collector current:

$$P_{ac} = \frac{E_{pp}I_{pp}}{8} = \frac{20\,(18.2 \times 10^{-3})}{8} = 45.5 \text{ milliwatts.}$$

Since the maximum d-c power is $E_c I_c = 11 \times 9 \times 10^{-3} = 99$ milliwatts, the efficiency becomes $\frac{P_{ac}}{E_c I_c} \times 100 = \frac{45.5}{99} \times 100 = 46\%$.

The maximum a-c signal current can be taken directly from the characteristic curve. In this case, $i_b = 600 \ \mu$a peak-to-peak. (The stabilizing resistors have been omitted to simplify the illustration.)

Push-Pull Operation. Whenever possible, transistor power amplifiers should be operated as push-pull stages. Push-pull operation has several desirable features, including the elimination of the even-order

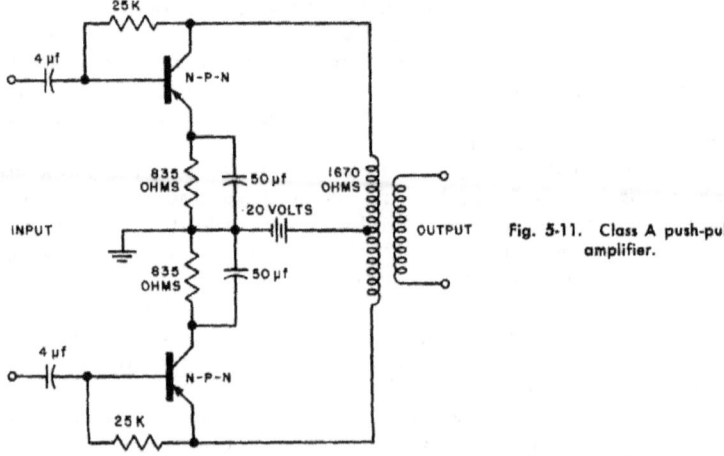

Fig. 5-11. Class A push-pull amplifier.

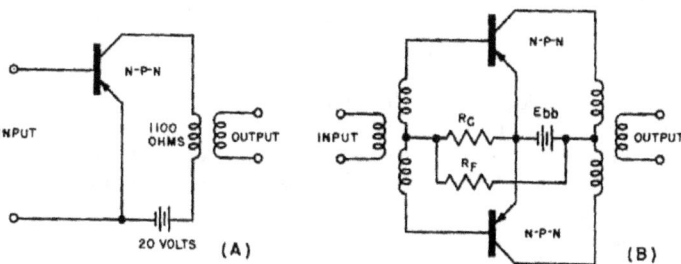

Fig. 5-12. (A) Class B circuit (constant voltage). (B) Class B push-pull operation.

harmonics and the d-c component in the load. The first factor is particularly fortunate, insofar as transistor applications are concerned. It was noted previously that operation at high values of collector current introduces a distortion due to crowding of the collector current-voltage lines. Thus, for a given value of allowable distortion, push-pull operation will allow the transistors to be driven into the higher I_c regions. In turn, each transistor delivers more power to the load than when it is connected for single-ended operation.

The operating point, load, and biasing resistors for the Class A push-pull stage are determined for each transistor exactly as if it were a single-ended type. A typical push-pull transistor amplifier is illustrated in Fig. 5-11, based on the same transistor characteristics used previously. The separate biasing arrangement indicated in this illustration permits a more exact match of the transistor characteristics. Notice that the load is twice the value computed for the single-ended stage.

Class B Transistor Amplifiers

Basic Operation — Quiescent Point. While the efficiency of a Class A amplifier is good under operating conditions, the collector dissipation is approximately the same whether or not a signal is applied. Its efficiency for intermittent or standby operation is poor. For standby operation, as in the case of the vacuum tube, Class B operation is preferred, and the operating point of a Class B transistor amplifier should be on the $E_c = 0$ line. This bias condition, however, would require an extremely high resistance in series with the battery. Thus most of the available supply power would be lost in the series resistor, the only function of which was to convert the voltage source into a current source. As an alternate method, a constant voltage battery is used. This sets the d-c operating point at the collector voltage $E_c = E_{bb}$ on the $I_c = 0$ axis. Figure 5-12 (A) shows a typical Class B transistor amplifier with a constant voltage source, using the same transistor as in previous calculations.

Push-Pull Circuitry. Two Class B amplifiers connected as a push-pull stage, using two of the circuits illustrated in Fig. 5-12 (A), will not operate. One transistor will always be biased in the reverse direction by the input signal, thereby causing its input resistance to become very high. This condition can be eliminated by using a center-tapped input transformer and connecting the center tap to the common emitter electrodes. This circuit is characterized by a distorted output wave. The distortion is particularly evident when the signal generator resistance is low. However, the distortion can be reduced within limits by introducing base bias into the circuit.

Figure 5-12 (B) illustrates one possible form of this latter arrangement. The value of the base bias resistor R_F for minimum cross-over distortion can be determined by the conventional graphic methods of vacuum-tube Class B push-pull amplifiers when using the composite transistor characteristics. The proper bias setting may be determined experimentally by direct measurement with an oscilloscope or a distortion meter. If the experimental method is used, care must be taken to avoid setting the base bias too high. This would cause a relatively high quiescent d-c collector current to flow, and the circuit would perform in a manner similar to that of a Class AB amplifier in vacuum-tube circuits. Resistor R_c may be a thermistor or some other temperature sensitive device. R_c is usually required in stages, subject to large changes in temperature to prevent excessive variation in the collector d-c operating point.

Another arrangement for a transistor push-pull Class B stage is illustrated in Fig. 5-13 (A). This circuit permits the elimination of the input transformers. The diodes D_1 and D_2 prevent each transistor from

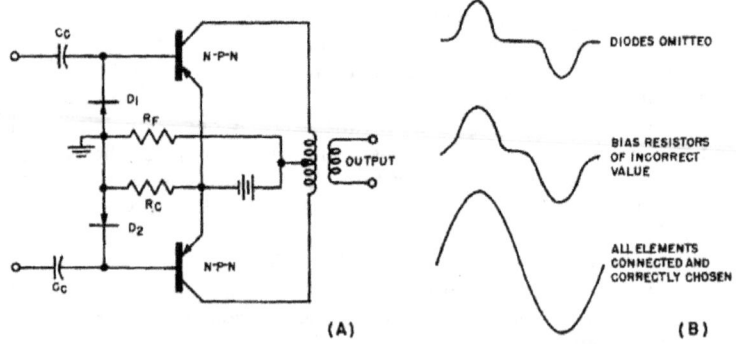

Fig. 5-13. (A) Class B push-pull operation without input transformer.
(B) Output waveforms.

cutting off when it is biased in the negative (reverse bias) direction by the input signal, since the diodes effectively short out the signal-induced bias. The point at which this bypass action occurs is determined by the bias due to resistors R_F and R_c. These resistors also furnish base bias to the transistors to minimize cross-over distortion. Figure 5-13 (B) illustrates the effect of diodes and bias resistors on distortion of the output signal.

The detailed operating characteristics of a Class B transistor push-pull amplifier are determined by the same methods used in similar vacuum tube circuits. The approximate values of the major characteristics can be calculated as illustrated in the following example: Assume that the transistors to be used in the Class B push-pull circuit have a maximum collector dissipation rating of 100 milliwatts, and assume that a battery $E_{bb} = 10$ volts is specified. The collector dissipation P_c in each transistor is approximately $\dfrac{E_{bb}I_{pc}}{8}$ where I_{pc} is the peak collector current. Then $I_{pc} = \dfrac{8P_c}{E_{bb}} = \dfrac{8(100)}{10} = 80$ milliamperes. The required load for maximum power output is: $R_L = \dfrac{4E_{bb}}{I_{pc}} = \dfrac{4(10)}{.08} = 500$ ohms; and the power output is approximately $\dfrac{E_{bb}I_{pc}}{2} = \dfrac{10(80)}{2} = 400$ milli-watts, or four times the maximum collector dissipation of each transistor.

Phase Inverters

Function. Transistor push-pull amplifiers, like their vacuum-tube counterparts, require the use of a phase inverter to supply the required balanced signal input. Transistor inverters are more complicated than conventional vacuum-tube types in that they must provide a balanced current, rather than a balanced voltage, input signal. However, the principles of operation are essentially the same.

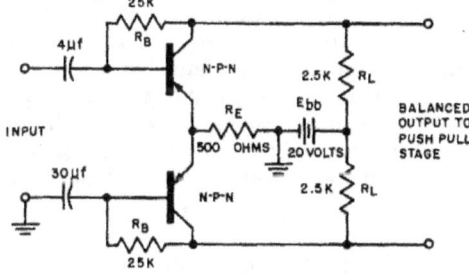

Fig. 5-14. Transistor phase inverter.

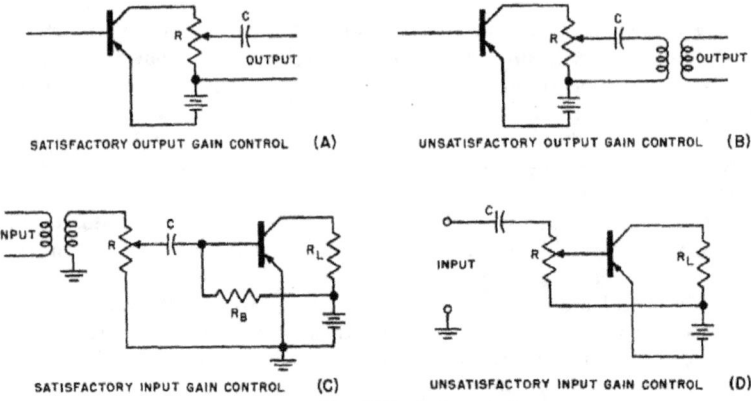

SATISFACTORY OUTPUT GAIN CONTROL   (A)   UNSATISFACTORY OUTPUT GAIN CONTROL   (B)

SATISFACTORY INPUT GAIN CONTROL   (C)   UNSATISFACTORY INPUT GAIN CONTROL   (D)

Fig. 5-15. Gain controls.

Typical Circuit. Figure 5-14 illustrates the basic circuit of a transistor phase inverter, which provides a reasonably well-balanced output. The basic operation is as follows: the upper transistor operates as a conventional grounded emitter amplifier except that the emitter is grounded through the parallel circuit, consisting of the lower transistor emitter-base path and resistor R_E. The emitter-base path has a low resistance, less than 50 ohms, so that practically all of the a-c emitter current of the top transistor flows through this path. Since the emitter current value for each transistor is the same, the collector currents are also equal if the current gains from emitter to collector are equal. For proper operation, the load resistances should be small compared to the output resistances of the transistors, and the emitter-to-collector current gains should be well matched. For the circuit illustrated, the output resistance of each transistor is the collector resistance shunted by R_B. Since r_c is much greater than R_B, the output resistance is equal to R_B. Thus R_B should be about ten times R_L. It is not necessary for the current gains to be exactly matched. Values which fall in the range of .92 to .97 are usually satisfactory. R_L and R_B are selected to provide the operating biases, which in this case are $E_c = 10$ volts, $I_c = 4$ ma, and $I_b = 400$ μa. The value of R_E is particularly important. It must be large compared to the emitter-to-base resistance path of the lower transistor; if it is not, an appreciable portion of the a-c signal will be shunted through R_E and the currents in the emitters will not be equal. In general, a value of R_E that is ten times the emitter-to-base circuit resistance is satisfactory.

Transistor Gain Controls

Despite the relatively low gain of transistor amplifiers, a gain control is frequently necessary to compensate for changes in the input sig-

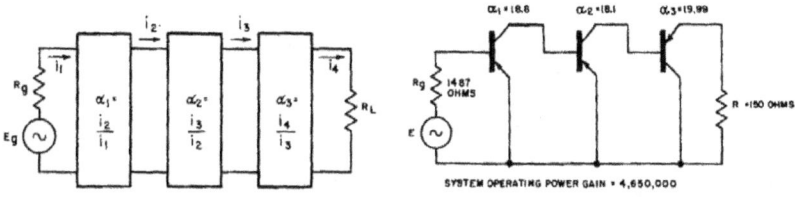

Fig. 5-16. Block schematic of cascade operation

Fig. 5-17. Calculated three-stage cascade.

nal, the ambient noise level, and other variations. The design of volume controls for transistor circuits is not a difficult problem if the fact that transistors are current operated devices is kept in mind. Figure 5-15 (A) illustrates one possible form of output gain control in a R-C coupled stage. In this circuit, the output potentiometer sets both the collector d-c operating point and the level of the output signal. The coupling capacitor blocks d-c current from flowing into the load. The value of this capacitor must be large enough to pass the lowest frequency to be amplified. If the output load is a transformer, this same form of gain control is not satisfactory, since, as illustrated in Fig. 5-15 (B), the load impedance varies with the potentiometer setting. If the coupling capacitor is omitted, circuit operation is poorer because the volume control setting changes the d-c operating point.

Figure 5-15 (C) illustrates a satisfactory form of input volume control in a transformer coupled stage. The resistance of the potentiometer should be at least ten times the value of the secondary-winding impedance to make its loading effect negligible. The arrangement illustrated in Fig. 5-15 (D), however, is not satisfactory, because the base bias varies with changes in the volume control setting.

In multistage operation, the gain control may be located in the input or output circuit of any stage. It is usually desirable to place the control in the first stage if the signal amplitude is likely to vary appreciably. This arrangement helps to prevent the system from overloading on large signals.

Cascade Operation

Design Considerations — Overall Power Gain. In any given problem requiring more than one stage of amplification, several cascade arrangements are possible. This flexibility is a desirable design feature; however, it complicates the problem of selecting the best combination of the three general forms of transistor connections with respect to the input and output resistances, and to the required gain of the system. Every design is fixed to some extent by the function of the circuit. but the requirement for maximum gain is invariably included.

Figure 5-16 is the block schematic of a three stage circuit. It is evident from inspection that the overall current gain of the system is the product of the individual stage gains, thus $a = a_1 a^2 a_3$. The operating gain as defined in equation $3\text{-}45$ is

$$G = 4R_g R_L \left[\frac{r_{21}}{(R_g + r_{11})(R_L + r_{22}) - r_{12}r_{21}} \right]^2$$

which can be modified to

$$G = 4R_g R_L \left[\left(\frac{r_{21}}{R_L + r_{22}} \right) \frac{1}{R_g + r_{11} - \left(\frac{r_{12}r_{21}}{R_L + r_{22}} \right)} \right]^2$$

Since the current gain as defined in equation $3\text{-}8$ is: $a = \dfrac{r_{21}}{R_L + r_{22}}$, and the input resistance as defined by equation $3\text{-}13$ is $r_i = r_{11} - \left(\dfrac{r_{12}r_{21}}{r_{22} + R_L} \right)$, these values may be substituted in the operating gain equation, which then becomes

$$G = \frac{4R_g R_L a^2}{(R_g + r_i)^2} \qquad\qquad Eq.\ (5\text{-}1)\ \bullet$$

This is a useful form of the equation. For the cascade stages, illustrated in Fig. 5-16, the overall power gain based on equation $5\text{-}1$ can now be written as

$$G = 4 \underbrace{\left[\frac{R_g a_1^2}{(R_g + r_i)^2} \right]}_{1st\ Stage} \underbrace{a^2_2}_{2nd\ Stage} \underbrace{R_L a^2_3}_{3rd\ Stage}$$

On this basis, a cascade system has maximum gain when each of the stages is separately designed for a maximum value of its associated gain factors.

Selection of Stage Connection. The first stage requires that its gain factor $\dfrac{R_g a^2_1}{(R_g + r_i)^2}$ be as large as possible. The following general rules for this stage are based on an analysis of the gain factor vs R_g characteristic:

1. When R_g has a low value (0 to 500 ohms), use either the grounded base or the grounded emitter connection.
2. When R_g has an intermediate value (500 to 1,500 ohms), use the grounded emitter connection.
3. When R_g has a high value (over 1,500 ohms), use the grounded emitter or the grounded collector connection.

In the intermediate stage a_2^2 is made as large as possible. This requirement generally can be met by either the transistor grounded emitter or grounded collector connection. The intermediate stage equivalent load should be less than $(r_c - r_m)$. If r_c is nearly equal to r_m, the grounded collector should not be used. This equality would cause the input and output resistances of the stage to become independent of the values of the connecting circuits. (An analysis of this buffer effect was covered in the discussion of input and output resistance of the grounded collector stage in Chapter 4.)

It must be noted that the intermediate stage represented by the current gain a_2 in this discussion may actually consist of several intermediate stages having a total current gain equal to a_2. This analysis of the three-stage circuit of Fig. 5-14, therefore, is applicable to any number of cascaded stages.

In the final stage, the gain factor $R_L a_8^2$ is made as large as possible. The following general rules for this stage are based on the analysis of the gain factor vs R_L characteristic for the three basic transistor connections.

1. When R_L has a small value (0 to 10,000 ohms), use a grounded collector or a grounded emitter connection.
2. When R_L has an intermediate value (10,000 to 500,000 ohms), use a grounded emitter connection.
3. When R_L has a high value (over 500,000 ohms), use a grounded emitter or grounded base connection.

(The numerical values listed above apply to those junction transistors with characteristics similar to the Western Electric Type 1752 transistor; however, the general values can be extended on a relative basis to cover all types.)

Based on the foregoing rules, it might appear that the choice of the grounded emitter connection is the best under all conditions. However, specific design problems often dictate the use of grounded base and grounded collector circuits when the coupling network, biases, feedback, and other factors are taken into consideration.

Cascade Design. As an illustration of these principles, consider the design of a three-stage cascade system using the typical junction transistor with $r_e = 50$ ohms, $r_b = 500$ ohms, $r_c = 1,999,500$ ohms, and $r_m = 1,899,500$ ohms. Assume that R_g is adjustable but limited to low values. $R_L = 150$ ohms requires the use of the grounded emitter or grounded collector connections. Assume that other design factors limit the choice to the latter case. Then for the last stage:

$r_{11} = r_c + r_b = 1,999,500 + 500 = 2,000,000$ ohms;
$r_{12} = r_c - r_m = 1,999,500 - 1,899,500 = 100,000$ ohms;
$r_{21} = r_c = 1,999,500$ ohms;
$r_{22} = r_c + r_e - r_m = 1,999,500 + 50 - 1,899,500 = 100,050$ ohms.

The input resistance of the last stage (equation *3-13*) is expressed as:

$$r_i = r_{11} - \left(\frac{r_{12}r_{21}}{r_{22}+R_L}\right) = 2 \times 10^6 - \left[\frac{0.1 \times 10^6\,(1{,}999{,}500)}{100{,}050+150}\right] = 5{,}000 \text{ ohms,}$$

and the current gain (equation 3-8) is:

$$a_3 = \frac{r_{21}}{R_L+r_{22}} = \frac{1{,}999{,}500}{150+100{,}050} \cong 19.99$$

Since r_m is close to the value of r_c, the intermediate stage is restricted to the grounded emitter connection. For this stage:

$r_{11} = r_e + r_b = 50 + 500 = 550$ ohms;

$r_{12} = r_e = 50$ ohms;

$r_{21} = r_e - r_m = 50 - 1{,}899{,}500 = -1{,}899{,}450$ ohms; and

$r_{22} = r_e + r_c - r_m = 50 + 1{,}999{,}500 - 1{,}899{,}500 = 100{,}050$ ohms.

Since the input resistance of the last stage is the output resistance of the intermediate stage, $R_L = 5{,}000$ ohms. The input resistance of the intermediate stage is

$$r_i = r_{11} - \left(\frac{r_{12}r_{21}}{r_{22}+R_L}\right) = 550 - \left[\frac{50\,(-1{,}899{,}450)}{100{,}050+5{,}000}\right] = 1{,}455 \text{ ohms}$$

and the current gain is:

$$a_2 = \frac{r_{21}}{R_L+r_{22}} = \frac{-1{,}899{,}450}{5{,}000+100{,}050} = -18.1$$

Since a low value of R_g is specified, the first stage must use either the grounded emitter or the grounded base connection. The load of the first stage equals the input resistance of the intermediate stage and is a low value. Therefore, the best choice for the first stage is the grounded emitter connection. Since R_g was specified as being adjustable, its value will be made equal to the input resistance,

$$r_i = r_{11} - \left(\frac{r_{12}r_{21}}{r_{22}+R_L}\right) = 550 - \left[\frac{50\,(-1{,}899{,}450)}{100{,}050+1{,}455}\right] = 1{,}487 \text{ ohms.}$$

The current gain is: $a_1 = \dfrac{r_{21}}{r_{22}+R_L} = \dfrac{-1{,}899{,}450}{100{,}050+1{,}455} = -18.75$

The overall current gain of the cascaded system

$$a = a_1a_2a_3 = (-18.75)\,(-18.1)\,(19.99) = 6{,}780$$

The operating gain (equation 5-1) is

$$G = \frac{4R_gR_La^2}{(R_g+r_i)^2} = \frac{4\,(1487)\,(150)\,(6780)^2}{(1487+1487)^2} = 4{,}650{,}000.$$

The resulting cascade circuit is shown in Fig. 5-17. This circuit does not include biasing arrangements, coupling networks and feedback loops. The values of the elements necessary for introducing these requirements may be computed by the methods in preceding paragraphs.

The cascade system may be changed considerably by the addition of external resistance arms to the circuits. These have the effect of increasing the effective values of the transistor parameters. For example, consider the effect of adding a stabilizing resistor $R_E = 50$ ohms in series with the emitter arm of the input stage. The effective resistance of the emitter is now $r_e + R_E = 50 + 50 = 100$ ohms, and the general four-terminal parameters are now:

$r_{11} = r_e + R_E + r_b = 50 + 50 + 500 = 600$ ohms;

$r_{12} = r_e + R_E = 50 + 50 = 100$ ohms;

$r_{21} = r_e + R_E - r_m + 50 + 50 - 1,899,500 = 1,899,400$;

$r_{22} = r_e + R_E + r_c - r_m = 50 + 50 + 1,999,500 - 1,899,500 = 100,100$ ohms.

The input resistance

$$r_i = r_{11} - \frac{r_{12}r_{21}}{r_{22} + R_L} = 600 - \left(\frac{100\,(-1,899,400)}{100,100 + 1,455}\right) = 2,472 \text{ ohms}$$

and the current gain $a_1 = \frac{r_{21}}{r_{22} + R_L} = \frac{-1,899,400}{100,100 + 1,455} = -18.72$

The overall current gain $a = a_1 a_2 a_3 = -18.72\,(18.1)\,(19.99) = 6,770$

and the operating gain $G = \frac{4R_g R_L a^2}{(R_g + r_i)^2} = \frac{4\,(2472)\,(150)\,(6770)^2}{(2472 + 2472)^2} = 2,790,000$.

Thus, a simple change reduces the overall system gain by a factor of one-half. It is evident that even after the basic stage connections are fixed, a considerable variation in the cascade performance and resistance terminal characteristics can be attained by changes in the effective value of the transistor parameters.

Coupling and Decoupling Circuits. To obtain the absolute maximum gain from a cascaded system, image resistance matching between stages is required. The analysis and conditions for matching the three basic transistor connections are covered in Chapters 3 and 4. The stage

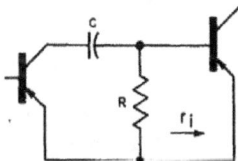

Fig. 5-18. (above). R-C interstage coupling; X_c less than r_i at lowest frequency to be amplified; R at least 10 times r_i.

Fig. 5-19 (right). Typical decoupling network.

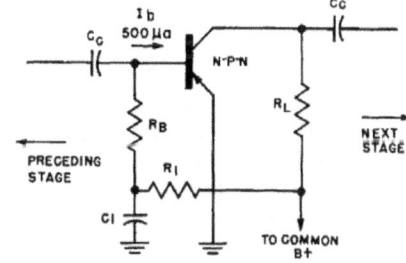

can be matched by interstage transformers. In i-f strips, transformer coupling is convenient and invariably used, because the transformers are also required for selectivity. In audio circuits, however, the increased gain due to the transformer is seldom worth its expense. In audio cascades, therefore, resistance-capacitance coupling is the most practical and economical choice. Figure 5-18 represents a typical R-C coupled stage. The capacitance must be large enough to pass the lowest frequency to be amplified. Its value can be computed as indicated in the preceding paragraphs dealing with single stage amplifiers. Resistor R must be large compared to the input resistance r_i. The interstage loss in gain is less than one db if R is chosen to be ten times as great as r_i.

When cascaded stages are connected to produce an overall gain of 60 db or more, consideration must be given to the addition of a decoupling circuit, as indicated by the combination R_1C_1, as shown in Fig. 5-19. Decoupling is required to prevent positive feedback through the battery resistance which is common to all the stages. High-gain transistor cascades almost always require a decoupling network, since even low values of battery resistance are significant when compared to the low input resistance of transistor stages. The product of R_1 and C_1 (time constant) should be equal to or greater than the inverse of the lowest frequency to be amplified by the stage. While this specified frequency sets the time constant, there are any number of combinations of C_1 and R_1 which can be used. In general, R_1 is made small enough so that it does not affect the supply voltage greatly, and at the same time is not made so low that a very high value of C_1 is required. The following example illustrates the calculation of the decoupling network: Suppose that for the circuit illustrated in Fig. 5-19, the d-c base bias $I_b = 500$ μa, and a drop of one-quarter of a volt in the battery supply through R_1 can be tolerated. The maximum value of R_1 equals the allowable voltage drop divided by the base current, $R_1 = \dfrac{.25}{500 \times 10^{-6}} =$ 500 ohms. If 100 cps is the lowest frequency to be passed, then $\dfrac{1}{f} = R_1C_1$ and $C_1 = \dfrac{1}{fR_1} = \dfrac{1}{100\,(500)} = 20$ μf. (In this equation, f is expressed in cycles per second, R_1 in ohms, and C_1 in farads.) The value of C_1 depends on the allowable voltage drop through R_1. If a larger drop is allowable the value of C_1 will decrease proportionately. In this example, assume that only a 10 μf capacitor is available, and that the maximum drop through R_1 can be increased. Then R_1, for the same cut-off frequency, equals $\dfrac{1}{fC_1} = \dfrac{1}{100 \times 10 \times 10^{-6}} = 1000$ ohms, and the voltage drop through R_1 equals $R_1I_b = 1000$ $(500 \times 10^{-6}) = 0.5$ volt. The base bias resistor now must be adjusted to compensate for the reduced value of the effective supply voltage. Thus

$$R_B = \frac{E_{bb} - R_1I_b}{I_b} = \frac{12 - 0.5}{500 \times 10^{-6}} = 23{,}000 \text{ ohms,}$$

as compared to the value (without decoupling),

$$R_B = \frac{E_{bb}}{I_b} = \frac{12}{500 \times 10^{-6}} = 24{,}000 \text{ ohms.}$$

In general then, when the value of the decoupling resistor is significant in comparison to the value of the bias resistor, R_B must be decreased by an amount equal to that of R_1 to maintain the specified d-c base current. In the form of an equation, this condition can be specified as:

$$\frac{E_{bb}}{I_b} = R_B + R_1$$

Figure 5-20 illustrates an experimental two-stage amplifier using grounded emitter circuits designed specifically to amplify the output of a 50 ohm dynamic microphone. The output terminates in a 600 ohm line. The overall gain of the system is 46 db.

Complementary-Symmetry Circuits

Basic Theory. The circuits discussed to this point can be used with either N-P-N or P-N-P transistors. It is necessary only that the battery supply is connected with the proper polarity. For other applications, it is possible and often very profitable to combine the two types of junction transistors into one circuit. This technique permits the design of many novel configurations that have no direct equivalent in vacuum tube circuits, since no one has yet invented a vacuum tube that emits positive particles from its cathode. Some of the characteristics of this unique property of transistors can be illustrated with the help of Fig. 5-21 (A), which is the composite curve of N-P-N and P-N-P units having identical characteristics except for polarity. (Practical circuits are never designed for an exact match, because of the expense of selection.) For

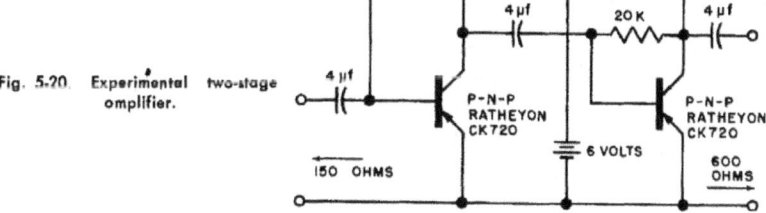

Fig. 5-20. Experimental two-stage amplifier.

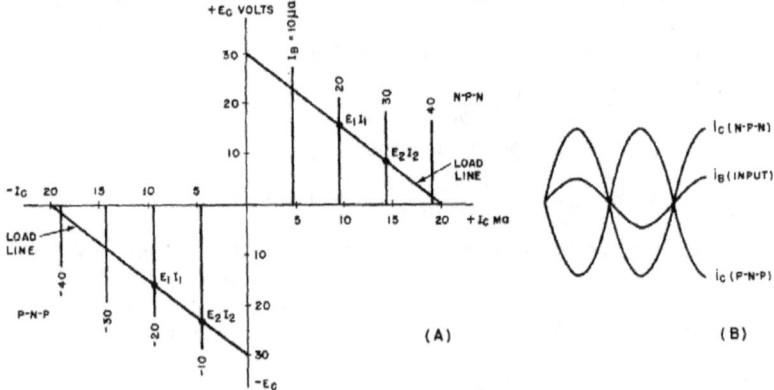

Fig. 5-21. (A) Composite characteristics for N-P-N and P-N-P transistors. (B) Waveforms of composite characteristics.

each operating point E_1I_1 in the N-P-N unit, there is an equivalent operating point $(-E_1)$ $(-I_1)$ for the P-N-P unit. These symmetrical properties offer innumerable possibilities in circuit applications. For example, if a peak signal current of 20 μa is applied to the base of each transistor simultaneously, the operating point of the N-P-N transistor has shifted to $E_2 = 8$ volts, $I_2 = 15$ ma at the instant that the input signal reaches a value of $+$ 10 μa. But at the same instant the operating point of the P-N-P unit is at $E_2 = -22$ volts, and $I_2 = -5$ ma. An increase in the base current of the N-P-N unit causes the collector current to increase; the same variation causes the collector current of the P-N-P unit to decrease. When the signal is reversed, the opposite effect occurs. The complete waveforms for this operation are shown in Fig. 5-21 (B). Since the output of the transistors are 180° out of phase, it appears that the N-P-N and P-N-P types will operate, with their input circuits in

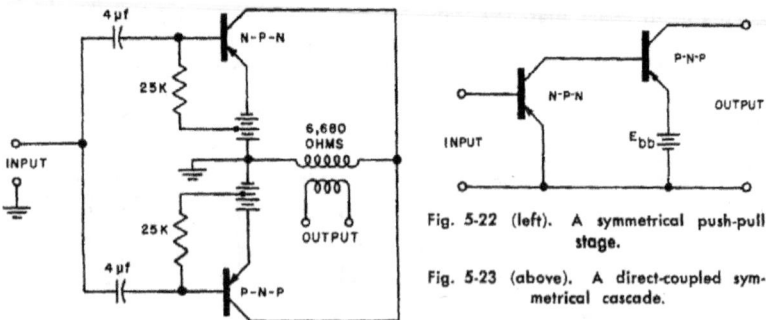

Fig. 5-22 (left). A symmetrical push-pull stage.

Fig. 5-23 (above). A direct-coupled symmetrical cascade.

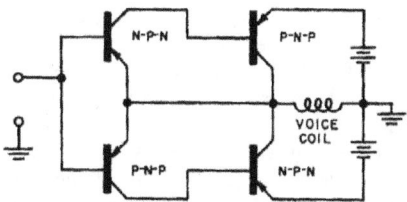

parallel, as a push-pull stage. Furthermore, due to the complementary action of the N-P-N and P-N-P types, the circuit does not require an input transformer or a phase inverter.

Symmetrical Push-Pull Operation. Figure 5-22 illustrates the basic symmetrical push-pull circuit with numerical values based on the same typical transistor characteristics used in previous examples. The operation of this circuit is the same as that of the transistor push-pull Class A amplifier that uses only one type of transistor. The circuit is capable of supplying a high voltage gain when operating into a high impedance load. The voltage gain of the circuit shown in Fig. 5-22 is in the order of 250 (48db). If the transistors are exactly symmetrical, the d-c collector currents supplied by each transistor cancel each other, and no d-c component flows in the load. The circuit is easily adaptable for direct connection to the voice coil of a speaker. Notice also that the same circuit can be modified by proper adjustment of the base bias for Class B push-pull operation.

Cascade Operation. One type of symmetrical circuit that proves very practical is the cascaded arrangement illustrated in Fig. 5-23. This tandem circuit represents the simplest possible cascade, since the only components of the system are the transistors and the battery supply. The gain per stage is low compared to the maximum available gain because of the mismatch existing between the stages. However, the reduced number of components and the simplicity of the design often outweighs this disadvantage.

A circuit which incorporates the major features of both push-pull and cascaded symmetrical configurations is shown in Fig. 5-24. This arrangement can serve as a single-ended power amplifier to feed a low impedance speaker from a relatively high resistance source. The two transistors in the output circuit are operated in the grounded emitter connection. Therefore, the phase of the input signal is reversed in going from base to collector. The base of the last stage is connected directly to the collector output circuit of the input stage. Since the signal also undergoes phase reversal in the first stage, the output of the transistors on each side of the load are in phase. The stability of this circuit is very high because it incorporates 100 percent degenerative feedback. The large amount of feedback keeps the distortion very low, and also allows the load to be very small. Since the circuit is in effect a two-stage Class B push-pull amplifier, the standby collector dissipation is negligible. The amplifier is capable of delivering a constant a-c output of about 400 milliwatts using transistors rated at 100 milliwatts. In intermittent short term operation, the same amplifier can deliver about a watt without damage to the transistors.

It is apparent that complementary-symmetry circuits offer considerable promise for further investigation. Their use in the field of high quality, low-cost portable audio systems is particularly attractive because the output can be fed directly into a voice coil, thus eliminating the expensive and often troublesome output transformer.

Chapter 6
TRANSISTOR OSCILLATORS

This chapter deals with the operation and circuitry of transistor oscillators. In general, these fall into two categories: the feedback (or vacuum tube equivalent) types, and the negative-resistance (or current multiplying) type. Transistor oscillators are capable of sine-wave generation by every mode of operation now feasible in vacuum-tube circuits, plus some additional novel modes. This chapter covers the capabilities of the transistor as an oscillator in basic rather than specific designs. A number of numerical examples and specific values are included to illustrate the fundamental concepts involved. An analysis of relaxation, frequency multiplication, frequency division, and triggering in the transistor is also included.

Feedback Oscillators

Transistor Hartley Oscillator. In the earlier chapters, it was shown that transistor properties, in every important respect, are equivalent to those of the vacuum tube. It is reasonable then to assume that any vacuum-tube oscillator configuration has an equivalent transistor circuit. For example, consider the vacuum-tube oscillator, illustrated in Fig. 6-1 (A), which represents one form of Hartley oscillator. Positive feedback is accomplished by arranging the resonant tank E to be common to both the input grid and output plate circuits. The equivalent transistor circuit using a grounded emitter connection is illustrated in Fig. 6-1 (B). Again, positive feedback is provided by placing the resonant tank so that it is common to both the input base and output collector circuits. If ground is removed from the emitter lead, and placed at the bottom of the tank circuit, the electrical operation of the oscillator is

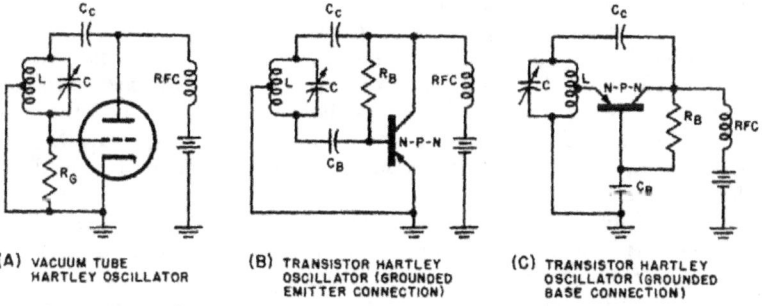

(A) VACUUM TUBE HARTLEY OSCILLATOR

(B) TRANSISTOR HARTLEY OSCILLATOR (GROUNDED EMITTER CONNECTION)

(C) TRANSISTOR HARTLEY OSCILLATOR (GROUNDED BASE CONNECTION)

Fig. 6-1. Vacuum tube and transistor Hartley oscillator circuits.

unchanged. Notice that when this circuit is rearranged as illustrated in Fig. 6-1 (C), it is now in the grounded-base connection. While the grid bias of the vacuum-tube oscillator in Fig. 6-1 (A) is regulated by the grid leak resistor R_G, the equivalent transistor base in Fig. 6-1 (C) is self-biased through resistor R_B. In all three circuits, the battery supply is decoupled by an R-F choke.

The major difference between the operation of the vacuum-tube Hartley oscillator and that employing a transistor lies in the loading effect of the emitter resistance on the tank coil. This resistance is reflected into the tank circuit and acts as an equivalent shunting resistance. The tank is also shunted by the collector resistance, and the equivalent shunt resistance of the resonant circuit becomes $R = \dfrac{\omega L}{Q}$. Oscillation starts when the equivalent shunt resistance of the tank is counterbalanced by the reflected negative-resistance of the emitter. The optimum tap point of the coil (as determined both mathematically and experimentally) is $T = \dfrac{a}{2}$, where T is the ratio of the feedback turns included in the emitter circuit to the total number of tank coil turns, and a is the emitter-to-collector current gain. Notice that when a approaches unity, the transistor oscillates at highest efficiency with a center-tapped tank coil. Under this condition the minimum allowable parallel resistance of the tank circuit is $R = \dfrac{4r_e}{a^2}$, which sets the Q of the circuit at $Q = \dfrac{R}{\omega L} = \dfrac{4r_e}{\omega L a^2}$, where $\omega = 2\pi f_0$, r_e is the transistor resistance, f_0 is the resonant frequency, and L is the inductance of the tank coil. The operating resonant frequency is always lower than the isolated resonant tank frequency, because of the change in effective value of inductance caused by the coil tap.

The disadvantages of tapping the coil can be avoided by using a direct feedback path from the resonant circuit to the input terminal. Figures 6-2 (A) and 6-2 (B) illustrate two such possible arrangements. In both examples, the feedback resistor R_F (a choke may be used) and the effective impedance of the resonant circuit form an a-c voltage divider. The value of R_F can be adjusted to obtain the required amount of feedback for sustained oscillation.

Transistor Clapp Oscillator. The transistor equivalent of the Clapp oscillator is illustrated in Fig. 6-3 (A). The operating frequency is set by the series resonant circuit in the collector circuit. Feedback is taken from the voltage divider consisting of capacitors C_1 and C_2. The upper frequency limit depends largely on the transistor in use, and can be increased

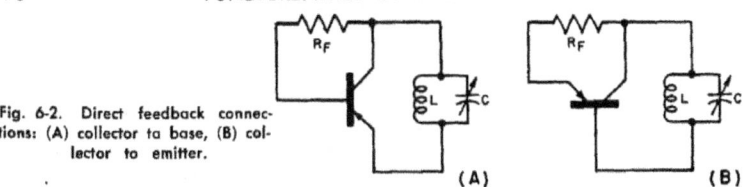

Fig. 6-2. Direct feedback connections: (A) collector to base, (B) collector to emitter.

(A)   (B)

considerably by careful selection of the unit. Upper frequencies as high as 3 mc can be attained using typical junction types in this basic circuit. The numerical values shown are based on the average of a group of Raytheon CK720 transistors. If crystal control is used, the frequency stability is improved, and there is also a considerable increase in the upper frequency limit of the oscillator. One method of achieving crystal control in this circuit is to replace the collector resonant circuit by a crystal.

Transistor Colpitts Oscillator. The transistor Colpitts oscillator is similar to the Clapp type except that the resonant load is a parallel arrangement in the collector circuit. Thus the circuit becomes voltage, rather than current, controlled. The feedback is again taken from a point between the two series capacitors connecting the collector to ground. The upper frequency limit for this oscillator is in the same range as that of the current-controlled Clapp arrangement. The typical values illustrated in Figure 6-3 (B) are again based on the average of a small group of Raytheon CK720 transistors.

In both cases, the parallel combination $C_B R_B$ provides the necessary emitter bias. This arrangement provides some degree of amplitude stability similar to the control provided by bypassed cathode or grid leak resistors in vacuum-tube oscillator circuits.

In servicing transistor oscillators, the emitter bias measured at the base end of the $C_B R_B$ combination is a useful indication of the signal

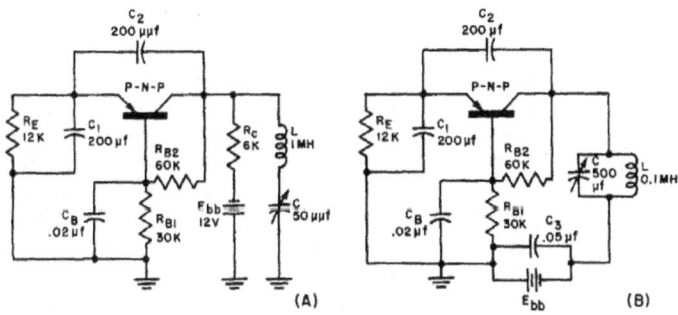

Fig. 6-3. (A) Transistor Clapp oscillator. (B) Transistor Colpitts oscillator.

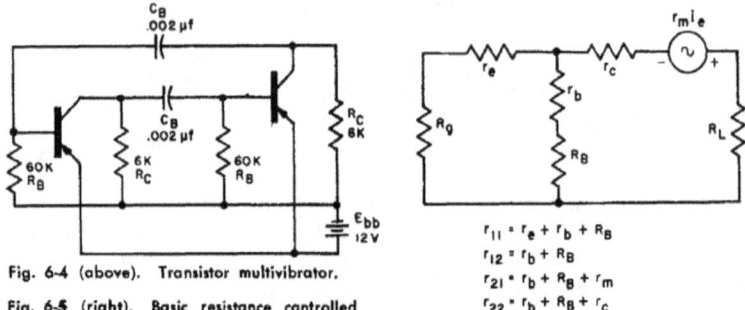

Fig. 6-4 (above). Transistor multivibrator.

Fig. 6-5 (right). Basic resistance controlled negative resistance circuit.

$$r_{11} = r_e + r_b + R_B$$
$$r_{12} = r_b + R_B$$
$$r_{21} = r_b + R_B + r_m$$
$$r_{22} = r_b + R_B + r_c$$

amplitude. In addition, the variation of the emitter bias over the frequency range indicates the relative uniformity of the signal output. Special care is necessary during these measurements to avoid affecting circuit operation. A vacuum-tube voltmeter may be used without causing additional loading. The use of a high resistance meter also minimizes that oscillator loading due to the stray reactance of the measuring probe. While direct current measurement is better, it requires disturbing the circuit wiring.

Transistor Multivibrator. Figure 6-4 illustrates the transistor equivalent of a basic multivibrator circuit. This configuration is generally useful in the frequency range of 5 to 15 kc. The parameter values shown are for an 8.33 kc oscillator which uses two Raytheon CK720 transistors. The frequency is determined by the $R_B C_B$ time constant. The value of R_B is limited to a maximum of about 200k ohms. C_B is limited to a minimum of .002 μf. The frequency stability is poor compared to the types previously discussed. The output collector waveform is almost a perfect square wave. The advantages of the transistor multivibrator are its simplicity and the small number of components required.

Negative-Resistance Oscillators

Conditions for Oscillation. The preceding paragraphs indicate that transistor oscillators can be designed as equivalents for all the known types of vacuum-tube oscillators that use an external feedback path. In addition, the unique property of a transistor that furnishes current gain can also be used to design many other novel types of oscillators. In the earlier chapters it was found that the point-contact transistor, by virtue of its ability to multiply the input current (r_m greater than r_c), is characterized by negative input and output resistances over part of its operating range. It is feasible, therefore, to use the point-contact transistor in this region to design oscillator circuits that do not require external feedback paths. As one engineer put it, "An oscillator is a poorly designed amplifier." This observation is particularly applicable in the

case of the negative-resistance oscillator. The conditional stability equation for a point contact transistor was specified in Chapter 4 as:
$(r_{11} + R_g) (R_L + r_{22}) - r_{12}r_{21}$ must be greater than zero. Thus for the transistor to be unstable, that is for it to exhibit negative resistance characteristics, requires:

$$(r_{11} + R_g) (R_L + r_{22}) - r_{12}r_{21} < 0 \qquad\qquad Eq. \ (6\text{-}1)$$

In general, external resistance can be added to any of the three electrode leads, as illustrated in Fig. 6-5. Substituting the transistor parameter values into equation 6-1 results in:

$$(r_e + r_b + R_B + R_g)(R_L + r_b + R_B + r_c) - (r_b + R_B)(r_b + R_B + r_m) < 0$$

Neglecting r_e and r_b as compared to R_B, r_c, and r_m, this becomes:
$$(R_B + R_g) (R_L + R_B + r_c) - R_B (R_B + r_m) < 0 \text{ and multiplying out}$$
$$R_B R_L + R_B{}^2 + R_B r_c + R_g R_L + R_g R_B + R_g r_c - R_B{}^2 - R_B r_m < 0$$
which becomes:

$$R_g (R_L + r_c) + R_B (R_L + R_g) - R_B (r_m - r_c) < 0$$

Notice that when r_m is less than r_c (as in the case of the junction transistor), the condition for oscillation cannot be satisfied. This re-emphasizes the fact that negative-resistance oscillators can only be designed using the point-contact transistor. Notice also in this equation that if both R_L and r_g are small compared to the value of $(r_m - r_c)$, the conditional equation is primarily controlled by the value of R_B. The higher the value of R_B, the more definite the instability. Furthermore, as the external collector and emitter resistances are increased in value, a higher resistance of R_B is required to assure circuit oscillation. The control of oscillation in negative-resistance transistor oscillators, then, is determined by the following three factors, either separately or in combination: the external resistance of the emitter lead (a low value favors oscillation), the external resistance of the base lead (a high value favors oscillation), and the external resistance of the collector lead (a low value favors oscillation).

Basic Operation. If the control of an oscillator can be maintained by simple high or low resistance values in the three transistor electrode arms, the substitution of series and parallel L-C resonant circuits in their place is a natural step. The insertion of a parallel resonant circuit in the base lead will cause the circuit to oscillate at the resonant frequency because of the tank's high impedance at resonance. On the other hand, placing a series L-C circuit in the emitter or collector arms will cause oscillation at the resonance frequency due to the tank's characteristic low impedance at that point. Fig. 6-6 illustrates the a-c equivalent circuit of a negative-resistance oscillator that includes all three methods of controlling oscillation. Since L-C resonant circuits produce sine waveforms, the oscillators using L-C resonant tanks are generally referred to as *sine-wave oscillators.*

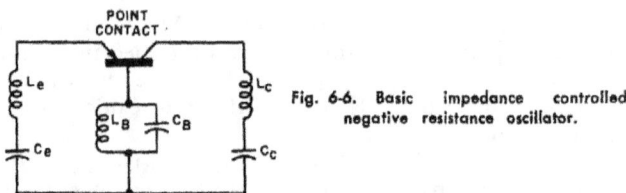

Fig. 6-6. Basic impedance controlled negative resistance oscillator.

The use of only the point-contact transistor for the negative-resistance oscillator is readily explained on an electronic basis. Assume that for the conventional grounded base connection, a disturbance or electrical charge of some sort causes an a-c emitter current to flow. This results in an amplified collector current $i_c = ai_e$ in the collector circuit. Since there is no phase inversion, the current flows through the base in phase with the emitter current. If the base resistance is large, the regenerative signal will be larger than the original signal. This increased current is again amplified, causing a greater collector current to flow, which again is fed back to the emitter, and so forth. In a short time, the current passes out of the linear dynamic operating range, and the circuit breaks into oscillation. The frequency of this oscillation is determined by the time constant of the circuit. In brief then, the point-contact transistor is capable of basic oscillation, without external feedback path, because of its ability to provide current gain and internal feedback path without phase reversal through the base lead.

General Types. Negative-resistance oscillators may be divided into two general classes: voltage controlled; and current controlled. The voltage-controlled oscillator is characterized by a high resistance load, and a low resistance power supply (constant voltage). The fundamental schematic of a typical oscillator of this type is illustrated in Fig. 6-7 (A).

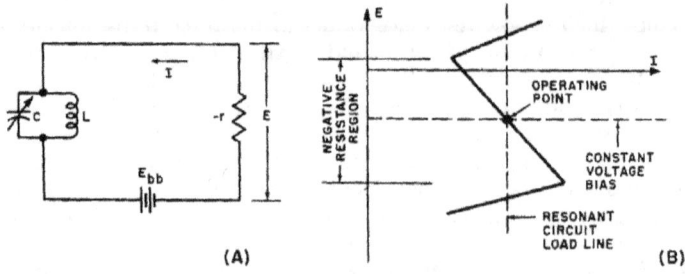

(A)

(B)

Fig. 6-7. (A) Voltage controlled negative resistance equivalent circuit. (B) Idealized current-voltage characteristic.

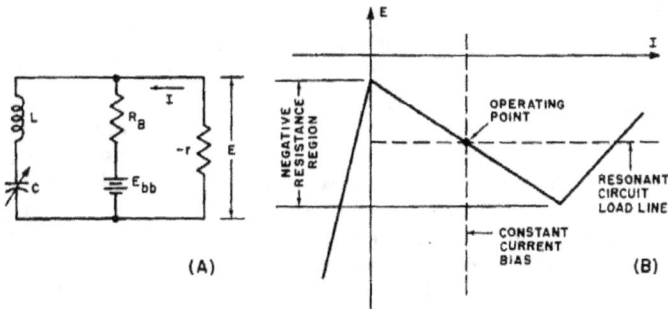

Fig. 6-8. (A) Current-controlled negative resistance equivalent circuit. (B) Idealized current-voltage characteristic.

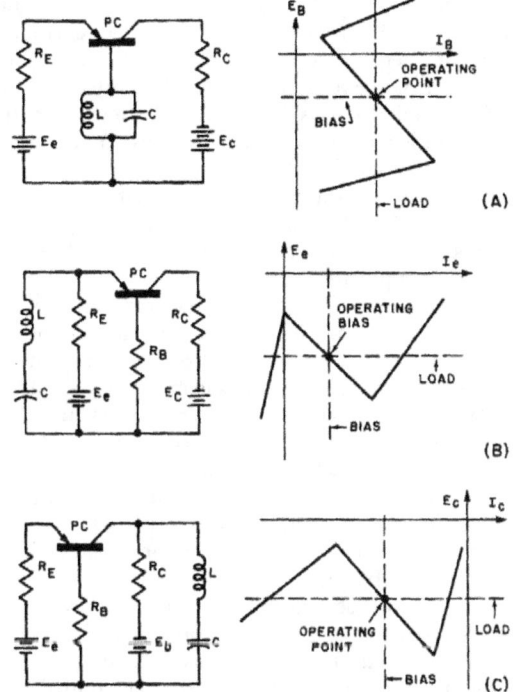

Fig. 6-9. (A) Base-controlled negative resistance oscillator and idealized characteristic. (B) Emitter-controlled negative resistance oscillator and idealized characteristic. (C) Collector-controlled negative resistance oscillator and idealized characteristic.

This oscillator is composed of three major parts: the resonant L-C circuit, the negative resistance of the oscillator, and the d-c supply voltage E_{bb}.

Figure 6-7 (B) represents the idealized current voltage characteristics of this oscillator. It is typical of the negative-resistance oscillator that the resistance remains negative only over a limited portion of its operating range. The bias is established somewhere in the middle of this useful section to guarantee oscillation. It is evident that a constant voltage bias is required. A remaining condition for sustained oscillation is that the resonant load have a higher absolute value than the negative resistance presented by the oscillator at the operating point. The parallel L-C circuit that approaches an infinite impedance at resonance, then, is ideal for this purpose.

The current-controlled type is shown in Fig. 6-8 (A). This oscillator is characterized by a low a-c load and a high d-c power source (constant current). Figure 6-8 (B) represents the idealized current-voltage characteristics for this negative-resistance oscillator. As in the voltage-controlled type, the negative-resistance region is limited to a section of the operating range, and the bias is established somewhere in the middle of this negative-resistance region using a constant current source. The last condition to be satisfied for sustained oscillation is that the a-c load of the resonant circuit must be less than the absolute value of negative resistance of the oscillator at the operating point. The series L-C circuit, the resonant impedance of which is close to zero, is the ideal load for this application.

Sine-wave Oscillators. These principles can now be applied to the three basic methods of controlling oscillation in the point-contact transistor: the insertion of low impedance loads in the emitter or collector circuits (current control), or the insertion of a high impedance load in the base lead (voltage control). Figure 6-9 (A) illustrates the basic base-controlled oscillator and its idealized current-voltage characteristics. This circuit is the most often used because it offers the best possibilities of the three types. Its main advantages are that it employs a constant voltage source (the easiest type to design), and that the regenerative feedback is through the resonant tank in the base lead. This latter feature assures frequency stability, because maximum feedback occurs at the resonant frequency of the tank circuit. The effect of the internal base resistance is negligible due to the extremely high value of the parallel circuit at resonance in comparison to r_b.

Figure 6-9 (B) represents the basic emitter controlled negative- resistance oscillator and its idealized current-voltage characteristics. Fig. 6-9 (C) is the basic collector controlled type. The fundamental operation of both is essentially the same. Oscillation occurs at the series resonant frequency of the L-C combination because at this point the

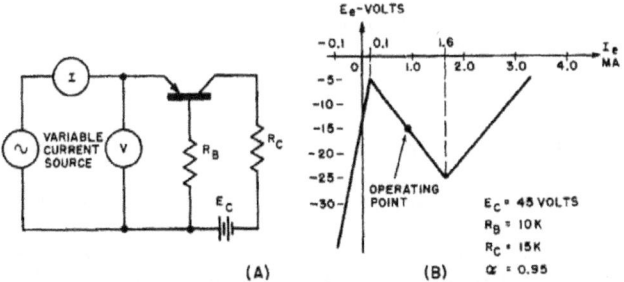

Fig. 6-10. (A) Basic measuring circuit for obtaining negative resistance characteristics. (B) Typical negative resistance characteristic.

effective resistance in either the emitter or collector arm is at minimum.

The base resistance must be large enough to furnish positive feedback in order to sustain oscillation. The base resistance r_b is generally large enough to cause instability when either the emitter or collector is shorted to ground, on the basis of equation 6-1. In practical circuits, however, r_b alone is rarely enough for dependable operation. An external resistor R_B equal to at least 2,000 ohms is generally added.

Negative Characteristic Measurements. The characteristics of the three basic negative-resistance connections are not generally supplied by the manufacturer. These, however, may be obtained by a point plot. This is not too arduous a task since the curves are reasonably linear and the changeover points are well defined. For most purposes it is sufficiently accurate to insert a sweep signal into the controlled electrode and observe the response on an oscilloscope. Figure 6-10 (A) illustrates the basic measuring circuit for this application when the transistor is in the emitter controlled connection. A typical resulting $E_e - I_e$ characteristic is shown in Fig. 6-10 (B).

The measuring circuit is easily modified for application to the base or collector controlled type. The plotted curve is similar to those illustrated in Figs. 6-9 (A) and 6-9 (C).

Bias Selection. It can be shown mathematically that the condition for locating the operating point in the center of the negative-resistance region is: $E_e(2aR_c + R_E) = E_cR_E$. This relationship indicates that the extent of the negative-resistance range depends upon the bias batteries and the values of R_E and R_c. The emitter-to-collector current gain a is, of course, fixed for a given transistor. For the characteristic in Fig. 6-10 (B), then, all the parameters are specified with the exception of E_e and R_E. Notice, however, that these quantities are related to the value of d-c emitter current bias I_e that is required to establish a d-c operating point in the center of the negative resistance region. This condition is: $E_e = R_E I_e$.

The two conditional equations can be combined to evaluate R_E in terms of known quantities:

$$R_E = \frac{E_c}{I_e} - 2_a R_c \qquad\qquad Eq.\ (6\text{-}2)$$

Since R_E also equals $\dfrac{E_e}{I_c}$, equation 6-2 limits the value of the emitter bias battery to less than that of E_c. The limiting value of $E_e = E_c$ is reached when $R_c = 0$.

As a numerical example, if the values associated with Fig. 6-10 (B) are used so that the bias current at the center of the negative resistance region is $I_e = 0.75$ ma, then

$$R_E = \frac{E_c}{I_e} - \left(2_a R_c\right) = \frac{45}{.75 \times 10^{-3}} - \left[2\,(.95)\,15 \times 10^3\right] = 31{,}500 \text{ ohms,}$$

and $E_e = I_e R_E = .75 \times 10^{-3} \times 31.5 \times 10^3 = 23.6$ volts.

The negative resistance of the oscillator is equal to the slope of the characteristics in that region. Then

$$r = \frac{E_{e\ (max)} - E_{e\ (min)}}{I_{e\ (max)} - I_{e\ (min)}} = \frac{(-25) - (-5)}{(1.6 - 0.1) \times 10^{-3}} = -\,13{,}300 \text{ ohms}$$

This value defines the maximum limit of the impedance of the L-C series emitter circuit at resonance.

Oscillator Stabilization. The generated signal of the sine-wave oscillator becomes badly distorted when the dynamic operating range of the circuit exceeds the negative-resistance region; excessive and uncontrolled distortion causes frequency instability. Obviously, the reduction of the harmonic content to a minimum is particularly important in those applications that require a stable and pure sine wave. But even in those cases where a high harmonic content is desirable, steps are necessary

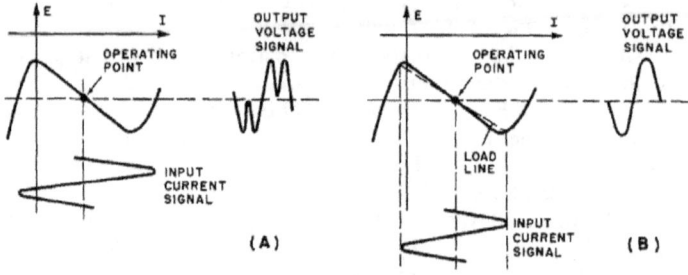

Fig. 6-11. Effect of a-c load on harmonic content: (A) idealized a-c resonant load; (B) A-c load slightly less than negative resistance of characteristic.

to keep the harmonic content of the signal constant to insure frequency stability of the oscillator.

The value of the a-c load impedance has a large effect on the amount of harmonic distortion in the signal. This effect is illustrated in Fig. 6-11 for the emitter controlled type. Figure 6-11 (A) illustrates the distortion in the voltage waveform when a sine wave of current is generated in an ideal series L-C load having zero impedance at resonance. Figure 6-11 (B) illustrates how the distortion is reduced to a satisfactory level by increasing the resonant impedance of the load. The increased a-c load effectively limits the dynamic range of the oscillator to the negative-resistance region. Thus, when a current-controlled oscillator is required to operate with a low harmonic content, the a-c load impedance should be chosen to be slightly less than the absolute value of the resistance determined by the slope of the negative-resistance characteristic. This same condition applies when the oscillator is collector controlled. A similar situation exists in the base-controlled negative-resistance oscillator except that, since this is a voltage-controlled oscillator, the distortion occurs in the current waveform. In this latter circuit, low distortion operation is attained by reducing the value of the resonant impedance so that it is slightly greater than the negative-resistance slope of the characteristic curve.

The operating point must be stabilized in the center of the negative-resistance region in order to avoid distortion from unequal positive and negative signal amplitudes. When the external resistances are fixed, the main causes of operating point shifts are changes in the bias supplies. An effective method of stabilization is the use of one supply battery for both the emitter and collector bias. This assures that the ratio $\dfrac{E_c}{E_e}$ will remain constant in spite of variation in the battery potential.

Increasing the resonant impedance of a series resonant arm is accomplished by selecting a higher resistance inductor, or by increasing the value of the series resistor in the emitter or collector lead. Decreasing the resonant impedance of the base controlled tank circuit is not as simple. A reduction of the tank Q will, of course, decrease the resonant impedance, but a low Q tank tends to promote frequency instability. A more satisfactory method of decreasing the impedance is to tap the base lead at some point in the tank coil. This permits the retention of a high Q tank, and, at the same time, reduces the effective impedance connected in the base lead. In addition, this connection helps to reduce the effects of internal transistor reactances on the operating frequency.

These internal reactances, primarily caused by junction capacitances, are particularly troublesome because their values do not remain

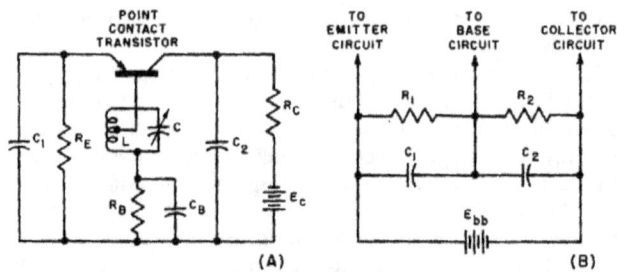

Fig. 6-12. (A) Stabilized base-controlled high-frequency oscillator. (B) Alternate method of providing common bias supply.

constant with changes in temperature and changes in operating currents and voltages. However, loose coupling between the tank and the base circuit minimizes the effect of internal transistor reactance. While this reduces the available power of the oscillator, the sacrifice of power for stable operation is generally justified. The level of the signal can always be increased by a stage or two of amplification.

Amplitude stability in negative-resistance oscillators is generally accomplished by incorporating some form of automatic bias control in the circuit. Sometimes the required amount of degenerative feedback is obtained through a non-linear resistor, placed in either the collector or emitter circuit. In this case, the main problem involves finding a non-linear element that is sensitive to the small current changes involved. Amplitude stability may also be obtained by a loosely coupled tank in the base-controlled oscillator, since it automatically decreases positive feedback at frequencies off resonance.

Stabilizing Circuitry. Figure 6-12 (A) illustrates one arrangement of a high-frequency base-controlled oscillator that incorporates the various stabilizing features discussed in the preceding paragraphs. C_1 and C_2 are phase compensating condensers. The base lead is connected to a tapped tank coil as a means of reducing the resonant impedance while maintaining a high Q tank. Bias stability is accomplished by using one common battery source. Notice also that positive emitter bias is supplied by the bypassed resistor R_B. Figure 6-12 (B) illustrates an alternate method of providing a constant collector-to-emitter bias ratio by means of a common battery supply. The advantage of this circuit is its design simplicity, since it is basically a voltage divider network. The values of C_1 and C_2 are not critical; they complete the a-c circuit between the collector, base, and emitter leads, and bypass the battery and bias divider network.

Except for the inductance and capacitance elements of the resonant network, the values of the external components in negative-resistance oscillators are not critical. The values of R_E and R_c should be large enough to limit their respective currents to safe values, but not so large that they cause excessive degeneration. The value of the base resistance R_B must be large enough to provide sufficient regeneration for sustained oscillation. Typical values for these parameters are: $R_E = 50$ to $2,000$ ohms; $R_c = 2,000$ to $10,000$ ohms; $R_B = 10,000$ to $20,000$ ohms.

Transistor Phase Shift

Contributing Factors. In general, transistor oscillators make use of their non-linear characteristics. While there has been considerable progress made in the mathematical analysis of non-linear circuits, particularly in the past few years, oscillator design is invariably based on the static characteristic curves. This is true since even the simplest mathematical approximations of non-linear operation are too involved for the average experimenter or engineer to handle.

When the operating frequency becomes more than 100 kc, the internal transistor parameters can no longer be considered as simple resistances. At this frequency, the values of the transistor reactive components become appreciable. In addition to the fixed-resonant circuit parameters, there are also stray reactances due to lead inductance, and others that have a considerable effect on the transistor characteristics. Static curves, then, are extremely useful to set bias points, and to approximate the negative-resistance range, optimum load, and waveshapes. However, circuit values based on the low-frequency transistor characteristics are not exact. The experimenter finds that every high-frequency transistor oscillator requires some readjustment for optimum operation.

Phase Shift and Feedback. One effect of the reactive components is to cause a phase shift between the input and output terminals. Phase shift reduces the in-phase component of the positive feedback signal. This is illustrated in Fig. 6-13 (A) where E_F is the feedback signal and ϕ

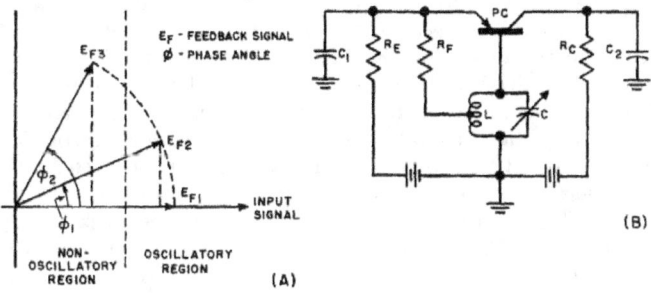

Fig. 6-13. (A) Effect of phase shift on feedback signal. (B) Base-controlled phase compensated oscillator.

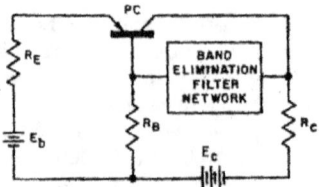

Fig. 6-14. Phase-shift oscillator.

is the phase angle between the input and output signals. E_{F1} represents the feedback amplitude at low frequencies when the reactive effects are negligible. As the operating frequency is increased, E_{F2} and the input signal are no longer in phase. Thus, only the in-phase component of E_F is useful for maintaining circuit oscillation. When the phase angle becomes so large that the in-phase component is less than the critical minimum required value, oscillation stops.

Phase Shift Compensation. The reduction of value of the in-phase feedback signal requires either an increase in feedback E_F or a form of phase compensation to decrease the angle ϕ. In the base-controlled oscillator, some phase shift compensation is provided by shunting either or both the emitter and collector electrodes to ground through a small capacitor (3 or 4 $\mu\mu f$). This simple modification usually doubles the upper frequency limit of a transistor.

One method of increasing the available feedback is to connect a resistor from the emitter to a tap point on the base tank coil. This provides regenerative voltage feedback to supplement the inherent current feedback of the circuit. The value of the resistor R_F is critical. The upper limit of the oscillator frequency drops as R_F is either increased or decreased from its critical value. For this reason, the feedback resistance is best determined on an experimental basis. Figure 6-13 (B) illustrates a basic oscillator incorporating these two methods of phase shift control and compensation.

Phase Shift Oscillator. One very stable negative-resistance oscillator is the phase-shift type illustrated in Fig. 6-14. This circuit is particularly useful in the audio range when a low distortion sine-wave signal is required. The resistances R_C, R_B, and R_E are determined by the condition for instability specified by equation *6-1*. The phase shift network used is a band-elimination filter at the desired operating frequency. At this frequency, the filter offers maximum attenuation (theoretically an open circuit). At any other frequency, the network attenuation decreases, thereby providing a degenerative feedback path into the base lead. This degeneration counteracts the positive feedback through the base resistor R_B. Thus, oscillation is favored only at the operating frequency, namely, the frequency eliminated by the phase shift network. If the network is designed for both phase reversal and minimum attenu-

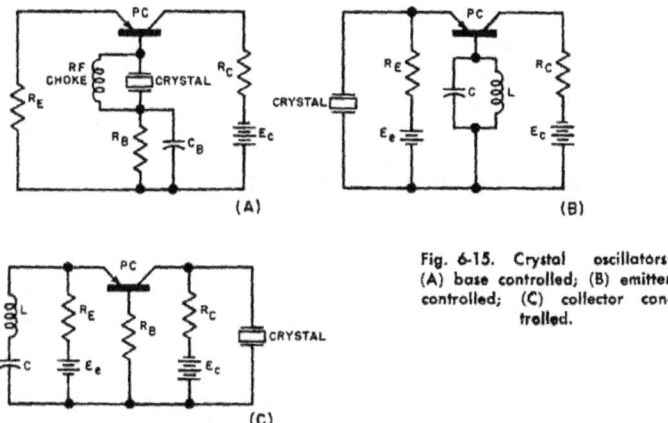

Fig. 6-15. Crystal oscillators: (A) base controlled; (B) emitter controlled; (C) collector controlled.

ation at the operating frequency, it will also be a useful oscillator. Under these conditions the network provides positive feedback into the base, which supplements the normal regenerative signal through the base resistor. The band-elimination filter oscillator is limited to the lower frequencies since proper operation depends on a zero phase shift through the network at the operating frequency.

Negative-Resistance Crystal Oscillators

Basic Types. The negative-resistance oscillator is easily adapted to crystal control, since crystals can operate as either series or parallel tuned circuits. Figure 6-15 (A) illustrates the basic circuit of the base-controlled crystal oscillator. The R-F choke which bypasses the crystal provides a d-c path to the base. A choke coil is used rather than a resistor for two important reasons: first, a resistor lowers the Q of the crystal; second, a resistor provides a positive feedback path for frequencies off resonance, thereby eliminating the major advantage of the base controlled circuit, namely, maximum regeneration at resonance, minimum regeneration off resonance. Since, in this case, the crystal is operated as a parallel resonant circuit, this oscillator is electrically equivalent to the base-controlled circuit illustrated in Fig. 6-12 (A).

Figures 6-15 (B) and 6-15 (C) represent the basic circuits of the emitter and collector-controlled crystal oscillators. The circuit shown in Fig. 6-15 (B) will operate satisfactorily if the base tank is replaced by a resistor. The inclusion of the tuned circuit, however, provides increased frequency stability and decreased harmonic distortion in the output signal. The series resonant circuit in the emitter arm of the collector-controlled oscillator illustrated in Fig. 6-15 (C) is added as a means of increasing the frequency stability. It can be replaced by a resistor.

Frequency Multiplication. Since the power handling capacity of the transistor is small, it can seldom provide enough energy to excite a crystal into oscillation at the higher frequencies. For this reason, high-frequency crystal-controlled oscillators usually incorporate some form of frequency multiplication. Figure 6-16 illustrates one basic circuit for a crystal-controlled frequency-multiplier oscillator. The emitter and base circuits in this base-controlled oscillator are conventional. The collector lead, however, contains a parallel resonant circuit tuned to the desired harmonic of the crystal fundamental frequency. At first glance it may appear that the inclusion of this network in the collector arm violates one of the fundamental requirements of negative-resistance oscillators, that is, the need for a low resistance collector circuit (equation 6-1). However, the collector tank is tuned to a harmonic of at least twice the fundamental frequency. Insofar as the fundamental crystal frequency is concerned, then, the collector tank is a low impedance. The tank offers a high impedance to the required harmonic, and consequently establishes a good feed point for this frequency into the output circuit.

Proper operation of the frequency-multiplier oscillator requires that the fundamental frequency be rich in harmonics, since low distortion contains little harmonic energy. The inherent non-linearity of negative-resistance oscillators [Figs. 6-9 (A), (B), and (C)], makes it easy to generate a distorted waveshape. This necessitates the use of a high impedance resonant circuit in the base-controlled oscillator, and the use of a low impedance circuit in the emitter or collector-controlled types. Tight coupling of the base tank also promotes increased harmonic generation, but this feature is generally unsatisfactory because of its adverse effect on frequency stability.

Relaxation Oscillators

Basic Characteristics and Operation. One of the most inviting applications of the negative-resistance oscillator is as a relaxation type, particularly since its power requirements are low. Transistor relaxation oscillators have almost limitless use where a complex waveform, pulse generation, triggered output or frequency division is required. Like the equivalent vacuum-tube types, the periodic operation of the transistor relaxation oscillator usually depends on a R-C or R-L combination for

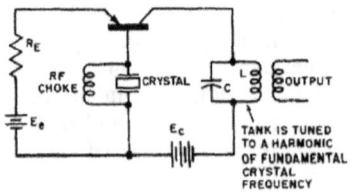

Fig. 6-16. Crystal-controlled frequency multiplier.

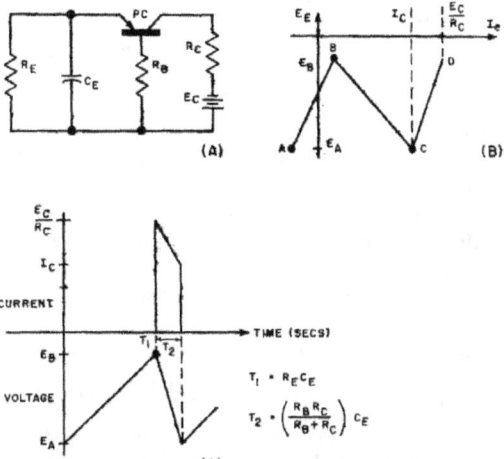

Fig. 6-17. (A) Basic emitter-controlled relaxation oscillator
with (B) idealized characteristic, and (C) waveforms.

the storage and release of signal energy. For this reason, they are characterized by abrupt changes from one operating point to another. This makes relaxation oscillators particularly useful for generating sawtooth waveforms.

Figure 6-17 represents the basic emitter-controlled relaxation oscillator and its idealized current-voltage characteristic. The location of the frequency-determining network in the emitter circuit provides the largest measure of control. This basic type, therefore, is the most useful. The fundamental operation is involved, but not difficult to understand. For simplicity, assume the operation starts at point A (Figure 6-17B). At this point the transistor is cut off, since the emitter is biased in the reverse direction $(-E_A)$. Because of this reverse bias, the input circuit offers a high resistance path. The charge on capacitor C_E (equals $-E_A$) has to leak off through R_E, and the rate of discharge is determined by the time constant $R_E C_E$.

When the voltage across the capacitor is reduced to $-E_B$, operation is at point B, which represents the point of transition from the cut-off to the negative-resistance region. The values of the emitter and collector resistances drop quickly to near zero, and the battery current is then limited only by the value of R_c. If the small effect of the saturation current I_{eo} is neglected, both the emitter and collector current increase from zero to $\dfrac{E_c}{R_c}$ almost instantaneously. In this instant, the operating point moves rapidly from point B through point C to point D. At the same time, the voltage across the capacitor starts to increase to its original value of $-E_A$. The rate is fixed by the time constant of C_E and the parallel equivalent of R_B and R_c. In the meantime, the emitter current decreases at the same rate, thereby moving the operating point back toward point C.

When the current reaches point C, operation passes from the saturation region to the negative-resistance region. Instability in this area causes the current to drop instantaneously to its value at point B. Because of this rapid drop, the condenser voltage does not change. The operating point returns to point A, and the condenser discharge action starts the cycle again.

Note that there are two time constants during a complete cycle. The first one $T_1 = R_E C_E$ controls the discharge rate of the condenser when operation moves from point A to point D. The second time constant $T_2 = \dfrac{R_B R_c}{R_B + R_c}$ (C_E) controls the charging rate when operation moves from point D to point A. The sawtooth voltage generated by this circuit is illustrated in Fig. 6-17 (C). The frequency of operation is approximately

$$F = \frac{1}{T_1 + T_2} = \frac{1}{C_E \left(R_E + \dfrac{R_B R_c}{R_B + R_c} \right)}$$

The frequency of the current wave is the same, but the waveform approximates a pulse, since the current only flows during the period when the condenser is charging (T_2). This simple oscillator, then, is useful as a voltage sawtooth or a current pulse generator.

The following problem will be used as a numerical example of basic relaxation oscillator design. Assume that a sawtooth voltage wave is required for use in a sweep circuit, and that the following characteristics are specified: frequency is 5 kc; the charging rate interval T_2 is limited to 10% of the total cycle; R_B is 2,000 ohms, required for sustained oscillation; E_c is fixed at 12 volts. The numerical values of the major operating points shown on Fig. 6-17 (B) are: for point A, $I_E = -0.1$ ma, $E_E = 10$ volts; for point B, $I_E = 0.01$ ma, $E_E = 2$ volts; for point C, $I_E = 3$ ma, $E_E = 10$ volts; and for point D, $I_E = 5$ ma, $E_E = 2$ volts. From the preceding analysis, $I_E = \dfrac{E_c}{R_c}$.

Thus at point D, $R_c = \dfrac{E_c}{I_c} = \dfrac{12}{5 \times 10^{-3}} = 2,400$ ohms

The overall time constant $T_o = T_1 + T_2 = \dfrac{1}{f} = \dfrac{1}{5 \times 10^3} = 200$ μ seconds, $T_2 = 10\% (T_o) = .10 (200) = 20$ μ seconds, and $T_1 = T_o - T_2 = 200 - 20 = 180$ μ seconds.

Since $T_2 = \dfrac{R_B R_c}{R_B + R_c}(C_E)$,

$$C_E = \dfrac{(R_B + R_c) T_2}{R_B R_c} = \dfrac{(2,000 + 2,400)\,20 \times 10^{-6}}{(2,000)\,(2,400)} = .018\ \mu f.$$

Since $T_1 = R_E C_E$, $R_E = \dfrac{T_1}{C_E} = \dfrac{180 \times 10^{-6}}{.018 \times 10^{-6}} = 10,000$ ohms

Base- and Collector-Controlled Oscillators. Base-controlled and collector-controlled relaxation oscillators are illustrated in Figs. 6-18 (A) and 6-18 (B). Both operate very much like the emitter-controlled type, and are analyzed on the basis of their respective operating characteristics, illustrated in Fig. 6-9 (A) and (C). The main difference is that the base-controlled type uses an inductance for the storage and release of circuit energy.

The fundamental difference between the sine wave oscillator and the relaxation oscillator is determined by which of the circuit parameters control the repetition rate. This, in turn, is determined by which has the lowest period of oscillation. For example, if in Fig. 6-12 (A) the time constant of the emitter network $C_1 R_E$ or the collector network $C_2 R_c$ is greater than that of the base L-C tank, the circuit becomes a relaxation oscillator. If a properly designed base-controlled high frequency sinusoidal oscillator suddenly switches to a different frequency and produces a distorted waveform, the trouble is most likely in the base resonant circuit.

While the R-C time constant of the collector- and emitter-controlled relaxation oscillator is fixed by the required operating frequency, the C to R ratio should be as high as possible. This causes minimum degeneration in the circuit, and, at the same time, increases the surge current handling capacity of the condenser. As before, the value of the base resistor R_B is determined by the amount of positive feedback required for sustained operation.

Self-Quenching Oscillator. The relaxation oscillator in combination with the regular base-controlled type can be used to form the self-quenching oscillator. Figure 6-12 (A) illustrates a self-quenching type if the value of either C_1 or C_2 is increased sufficiently to make the

Fig. 6-18. (A) Base-controlled relaxation oscillator. (B) Collector-controlled relaxation oscillator.

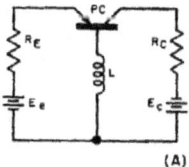

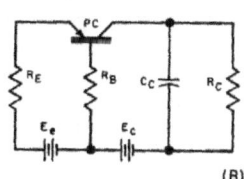

(A)   (B)

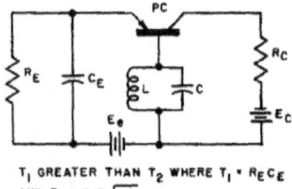

Fig. 6-19. Basic self-quenching oscillator.

T_1 GREATER THAN T_2 WHERE $T_1 = R_E C_E$
AND $T_2 = 2\pi\sqrt{LC}$

emitter or collector time constant appreciably greater than that of the L-C tank circuit. Figure 6-19 represents the basic self-quenching oscillator. Due to its time constant, the R-C emitter network has primary control of the circuit and produces the sawtooth voltage and pulsed current waveforms illustrated in Fig. 6-17 (C). The operation of the relaxation section of the circuit is independent of the base tank. The base network, however, depends entirely on the relaxation operation. Assume the cycle is moving in the charging direction (B of Fig. 6-17), operation from point C to point A. When the operation reaches the negative-resistance region where sufficient regenerative energy is supplied, the base tank oscillates at its resonant frequency. The amplitude of the resulting wave is small initially, but rises to a peak at the point when C_E starts its discharge cycle (B of Fig. 6-17), operation from point A to point D. The duration of the oscillation in the base tank is a function of the Q of the network, the amount of stored energy and the loading effect on the tank by the rest of the circuit. The relaxation or quench frequency in this case is $f_Q = \dfrac{1}{(R_E + R_c)\,C_E}$, while the resonant frequency of the tank is $f_T = \dfrac{1}{2\pi\sqrt{LC}}$. Notice that f_Q must be less than f_T for proper operation. The basic circuit becomes collector controlled if capacitor C_E is moved into the collector circuit. The circuit operation is exactly the same.

Synchronized Relaxation Oscillator. The operation of a synchronized relaxation oscillator is easily understood in view of the fundamentals of operation covered in the preceding paragraphs. The basic circuit is the same, but the relaxation frequency is made slightly less than the synchronizing frequency. Referring to Fig. 6-17 (B), assume that operation is moving from point A toward point B, and that a positive pulse, large enough to instantly move operation to point B, is applied to the emitter. The effect, as illustrated in Fig. 6-20 (A), is the same as decreasing the time constant $R_E C_E$, and the relaxation frequency becomes the same as that of the applied synchronizing pulse. The actual point at which the synchronizing signal arrives is not critical as long as the pulse amplitude is large enough to carry the opera-

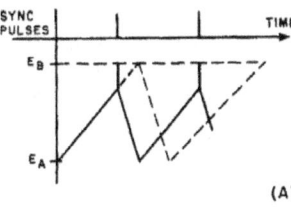 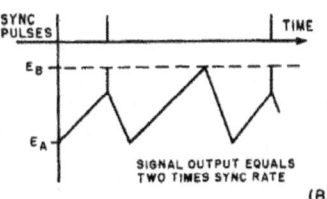

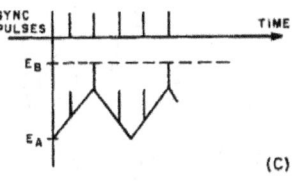

Fig. 6-20. Synchronized controlled (A) waveform, (B) frequency multiplication waveforms, (C) frequency divider waveforms.

tion into the negative-resistance region. Notice, however, that the magnitude of the sawtooth voltage is reduced by an amount equal to that of
the pulse. The dotted line represents the voltage waveform without
synchronization.

The synchronized oscillator can be used as a frequency multiplier.
Figure 6-20 (B) illustrates one application in which the input frequency
is approximately half that of the relaxation frequency. Any sub-multiple
of the normal rate will work. The chief disadvantage of this type of
operation is the lack of control over the frequency in the interval during synchronizing pulses.

Figure 6-20 (C) illustrates the application of the synchronized relaxation oscillator as a frequency divider. In this example, the input
frequency is three times that of the relaxation rate. As long as the
synchronizing rate is an integral multiple of the basic frequency, the
oscillator remains under control. Theoretically, any division ratio is
possible, but in practical circuits the ratio is limited by the non-linearity of the sawtooth wave near the critical voltage E_B. Consistent operation for division ratios up to approximately 10 to 1 can be easily attained. Ratios higher than these require critical design for reliable operation.

Negative synchronizing pulses can be used to operate the base or
collector-controlled oscillator types. The many ramifications of the basic
relaxation oscillator are too numerous to cover, but the experimenter
may find many useful applications for this circuit. If, for example, time
constants are inserted in both the emitter and collector circuits, the relaxation oscillator can be synchronized by a pulse applied to either

electrode. The circuit may also be biased in either the saturation or cut-off region, so that it remains non-oscillatory until pushed into the regenerative region by an external pulse. The last type falls under the general category of trigger circuits.

Trigger Circuits

The transistor oscillators considered to this point have one feature in common: the controlling electrode is biased in the negative resistance region. These types, whether synchronized, sinusoidal, or non-sinusoidal, come under the general classification of *astable operated*.

Triggered circuits, on the other hand, are biased in one of the stable regions and are non-oscillatory until the trigger pulse is applied. These types are classified as either *monostable operated* or *bistable operated* oscillators.

Monostable Operation. The basic monostable circuit is illustrated in Fig. 6-21 (A). The only difference between this circuit and the emitter-controlled relaxation oscillator illustrated in Fig. 6-17 (A) is the elimination of the emitter resistor R_E. Since this action removes the d-c emitter current bias ($I_E = 0$), the operating point shifts from the negative-resistance region (P_2) to the point intersection of the voltage axes at $I_E = 0 (P_1)$. This change is illustrated in Fig. 6-21 (B). The circuit is no longer capable of self-sustained oscillation since it is biased in the stable cut-off region. Now, if a pulse of sufficient magnitude (at least equal to I_P) is applied to the emitter, operation is forced into the regenerative region. The current jumps to its value at point D, and the negative charge on the emitter condenser starts to build up. When the charging current is reduced to its value at point C, operation again enters the regenerative region, and the current is quickly reduced to its value at A. The charge on the condenser gradually leaks off through the emitter base circuit ($r_e + r_b + R_B$) until the stable operating point P_1 is reached. The circuit is now ready for another trigger pulse.

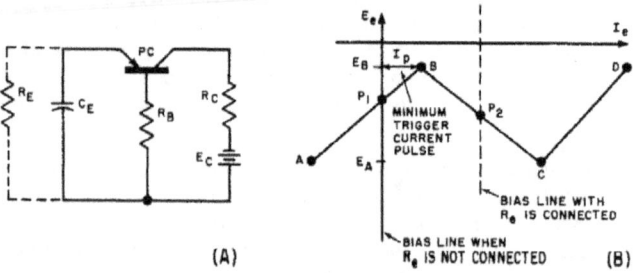

(A)   (B)

Fig. 6-21. (A) Basic monostable trigger circuit. (B) Idealized characteristic.

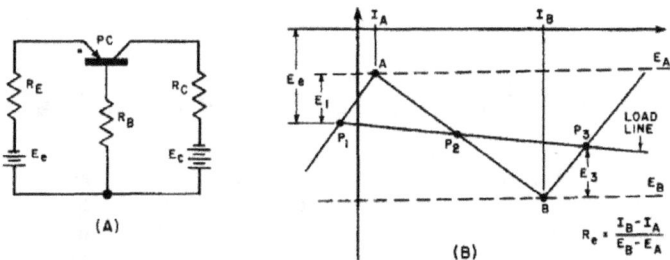

Fig. 6-22. (A) Basic bistable trigger circuit. (B) Idealized characteristics.

The emitter resistance is very high in the cut-off region due to the reverse bias. As a result, the same constant $C_E (r_e + r_b + R_B)$ is large compared to that of the relaxation type illustrated in Fig. 6-17 (A). This is the major factor limiting the repetition rate of the trigger pulse if sensitive operation is required.

Bistable Operation. Figure 6-22 (A) illustrates the basic bistable circuit. The fundamental requirement for this type of operation is that the load line intersects the characteristic curve once in each of the three operating regions. This automatically establishes three operation points: one in the unstable negative-resistance region; one in the saturation region; and one in the cut-off region. The last two points are stable, hence, circuit operation is properly defined as bistable. The operation shown in Figure (6-22 (B) is as follows: When operation is at point P_1, the circuit is stable, since the current is low; this is referred to as the *off-state*. If a positive pulse is now applied to the emitter, operation enters the regenerative region at point A. The operation swings rapidly to the saturation region where, at point P_3, the circuit is again stabilized. Since the current at this point has considerable magnitude, this is referred to as *on-state*. To move operation back into the off-state requires a negative trigger pulse whose magnitude is at least equal to E_3. This pulse moves operation back into the unstable negative-resistance region at point B, where it rapidly swings back to the stable off-state point P_1.

The value of R_E is selected to provide the three necessary operating points. It is not critical and may vary considerably but, in general, it should be fairly low. Notice that the potential of the emitter battery E_e fixes the location of P_1, which in turn determines the required value of the trigger pulse E_1. A low battery voltage, then, causes sensitive operation, since the triggering can be accomplished with a small pulse. A large value of E_e results in less sensitive but more reliable operation, since the circuit is less likely to be triggered by noise or other unwanted circuit disturbances. The final choice of both E_e and R_E should be based on the most sensitive combination providing reliability.

Chapter 7
TRANSISTOR HIGH FREQUENCY AND OTHER APPLICATIONS

The preceding chapters discussed the basic operation, circuitry, applications and limitations of the transistor. This chapter contains important miscellaneous considerations, including transistor operation at high frequencies, i-f and r-f amplifiers, limiters, mixers, handling techniques, hybrid parameters, and printed circuits.

The Transistor at High Frequencies

Transit Time, Dispersion Effect. In the earlier chapters it was noted that the low-frequency, small-signal parameters change as the operating frequency is increased appreciably above the audio range. Figure 7-1 illustrates the low-frequency equivalent circuit of the transistor including the collector junction capacitance C_c. At higher frequencies this equivalent circuit must be modified to include the effects of the current carriers' transit time on the transistor parameters. The transit time of the carriers (holes or electrons) is one of the major factors limiting the high frequency response of the transistor.

The movement of holes or electrons from the emitter through the base layer to the collector requires a short but finite time. In the transistor, as noted earlier, the electron does not have a clear and unimpeded path from emitter to collector. As a result, the transit time is not the same for all electrons injected into the emitter at any one instant. The effect of an identical transit time for all electrons would be a simple delay in the output compared to the input signal. Because the injected carriers do not all take the same path through the transistor body, those produced by a finite signal pulse at the emitter do not all arrive at the collector at the same time. The resulting difference is very small and is of no consequence in the audio frequency range. At the higher frequencies, however, this difference becomes a measurable part of the operating cycle, and causes a smearing or partial cancellation between the carriers. Figure 7-2 illustrates the dispersion effect in a tran-

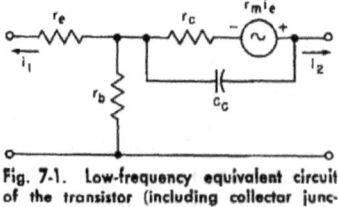

Fig. 7-1. Low-frequency equivalent circuit of the transistor (including collector junction capacitance).

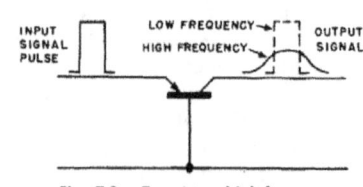

Fig. 7-2. Transistor high-frequency dispersion effect.

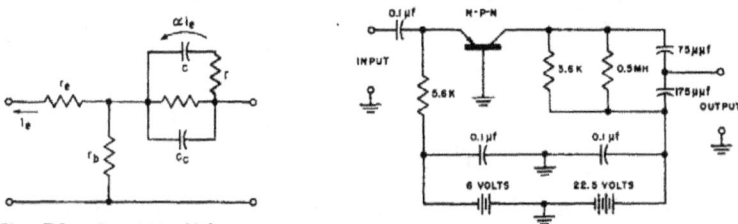

Fig. 7-3. Transistor high-frequency equivalent circuit.

Fig. 7-4. Typical transistor i-f amplifier.

sistor at high frequencies. Notice that, in addition to the increased period, the signal has also suffered a reduction in amplitude (the time delay results in a phase shift). The decrease in the output signal means a decrease in the current gain $a = \dfrac{i_c}{i_e}$. The degradation in frequency response becomes steadily worse as the operating frequency is increased, until eventually there is no relationship between the input and output waveforms (and no gain).

Another factor that limits the high frequency response of the transistor is the capacitive reactance of the emitter input circuit, which behaves as if r_e is shunted by a capacitor. This reactive parameter is reduced if the source impedance is made as low as possible. Since r_b is also effectively in series with the source, a good high frequency transistor must have a low base resistance. If the source impedance and base resistance are low, the upper frequency response limit is determined primarily by the collector junction capacitance and the variation in the current gain.

Alpha (a) *Current Frequency.* In view of these limitations, the basic circuit illustrated in Fig. 7-1 is not a useful approximation of transistor performance at high frequencies. To modify this circuit for accurate representation of high frequency equivalence requires that all of the internal parameters be specified in a complex form (magnitude and phase angle) as functions of the frequency. In most cases, however, it is sufficiently accurate to modify Fig. 7-1 to include only the variation of a with frequency, since few design problems justify the details required for exact equivalence. The variation in current gain can be satisfactorily approximated by the relationship:

$$a = \frac{a_1}{\sqrt{1 + \left(\dfrac{f}{f_c}\right)^2}}$$

where a is the current gain of the operating frequency f; a_1 is the low frequency current gain; and f_c is the frequency at which the current gain is 0.707 of its low frequency value (3 db down).

As a numerical example of the above, compute the current gain for a junction transistor having a low frequency current gain of $a_1 = 0.95$, an a cut-off frequency of $f_c = 10$ mc, and an operating frequency of 7.5 mc. Then

$$a = \frac{a_1}{\sqrt{1 + \left(\frac{f}{f_c}\right)^2}} = \frac{.95}{\sqrt{1 + \left(\frac{7.5}{10}\right)^2}} = 0.76$$

Including only the junction capacitance and variation in a in the low frequency circuit makes all the computed values far from exact. In addition to the capacitive reactance of the emitter, there is also considerable variation with frequency in the collector resistance and collector junction capacitance. The collector resistance r_c decreases rapidly for a ratio of $\frac{f}{f_c}$ greater than 0.15, falling to about 10% of its low frequency value at $\frac{f}{f_c} = 1$, and then remains at that value. The collector junction capacitance C_c also decreases as the operating frequency increases above an $\frac{f}{f_c}$ greater than 0.15, but does not decrease as rapidly as r_c. In a typical characteristic, C_c drops to approximately 75% of its low frequency value at $\frac{f}{f_c} = 1$ and to about 50% at $\frac{f}{f_c} = 10$, after which the curve levels out. Due to the coupling between the input and output circuits, $r_1 = r_{11} - \frac{r_{12}r_{21}}{r_{22} + R_L}$, the input impedance contains a reactive component beyond the emitter shunt capacitance. At the a cut-off frequency f_c, the reactive component is approximately equal to the resistive input component. This causes the input impedance to be inductive for the grounded base connection, and capacitive for the grounded emitter connection (due to phase reversal).

High Frequency Equivalent Circuit. Because of these factors, representation of the transistor high-frequency operation by any linear four-terminal equivalent network is at best a rough approximation over any substantial frequency range. This is especially true if the circuit is to be reasonably representative of the physics of the transistor, and if the number of circuit parameters are to be kept within reasonable limits. One form of equivalent circuit, suggested by Dr. W. F. Chow of the General Electric Company, has worked out well. This involves the insertion of a low pass R-C filter network in the low frequency circuit, derived for an equivalent current generator in the collector arm (ai_e). The modified equivalent circuit illustrated in Fig. 7-3 takes into account the variations of r_c and C_c with frequency. This circuit provides a fair representation of transistor performance through the range below the a cut-off frequency. If the operating frequency is greater than f_c, the low pass filter must be replaced with an R-C transmission line.

Frequency Comparison of Point-Contact and Junction Transistors. At this point, a brief explanation of why the point-contact transistor is capable of a higher operating frequency than the junction type is in order. The high frequency effects on the equivalent circuit parameters are essentially the same for both types. Actually, the major difference is in the mechanics of conduction.

Point-contact transit time is determined primarily by the field set up by the collector current. In equation form, $T = \dfrac{2\pi S^3}{3\mu\rho I_c}$ where S is the point spacing in centimeters, μ is the hole mobility in $\dfrac{cm/sec}{volts/cm}$, ρ is the germanium resistivity in ohm-cm, and I is the collector current in amperes. Typical transistive values are $S = .003$ cm, $\mu = 100\, \dfrac{cm/sec}{volts/cm}$ $\rho = 12$ ohm-cm, and $I_c = 3$ ma, for which $T = 1,570\ \mu\mu$secs. Ignoring all other factors, this limits the upper frequency response of the point-contact transistor to about 600 megacycles.

In the junction transistor, movement of the current carriers is primarily by diffusion, and is not appreciably affected by the electrode potential fields. In equation form $T = \dfrac{W^2}{D}$, where T is the diffusion time through the base layer, W is thickness of the base layer in centimeters, and D is the diffusion constant in cm^2/sec. Typical values for a P-N-P transistor are $W = 2 \times 10^{-3}$ cm and $D = 33\ cm^2/sec$ (for an N-P-N type, D is about 69 cm^2/sec), for which $T = 0.121\ \mu$secs. Ignoring all factors but the diffusion time, the upper frequency for this typical P-N-P type is approximately 8 mc, and for the N-P-N type about 16 mc.

High Frequency Circuits

I-F Amplifiers. In general, the upper frequency limit of the junction transistor is considerably lower than the limits of the point-contact type. On the other hand, the junction type has a lower noise factor, and better stability in some applications. These factors frequently make it advantageous to use the junction transistor in some high frequency applications even if an additional stage or two may be required.

Figure 7-4 illustrates one stable form of i-f amplifier stage using a WE 1752 N-P-N transistor. The operating frequency is 455 kc, and the gain is 18 db.

Due to the natural regenerative feedback path through the collector junction capacitance and the base resistance, and the close coupling between the input and output circuits, the circuit, when connected in tandem, is likely to oscillate unless the stage is carefully tuned. The

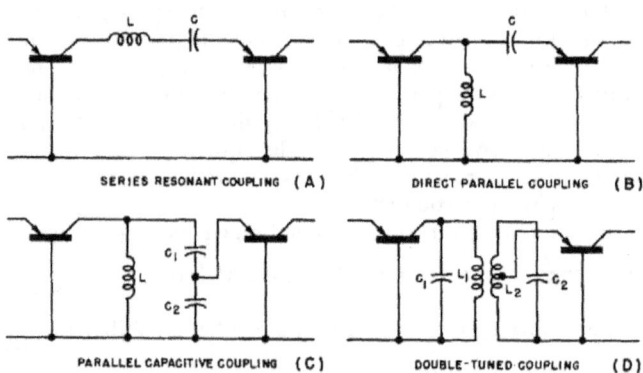

Fig. 7-5. Transistor i-f coupling networks.

alignment procedure is easiest if the last stage is tuned first. For an input resistance $R_g = 500$ ohms, the output resistance r_o averages 12,500 ohms, and C_c is about 15 $\mu\mu f$.

The cascading of transistor i-f is more complicated than that of vacuum tubes. The main contributing factors are the effect of the output load on the input impedance, and the effect of the generator impedance on the output impedance. These factors show up largely in the design of interstage coupling networks.

I-F Coupling Circuits. For interstage coupling, an i-f transistor amplifier may use a series resonant circuit such as that illustrated in Fig. 7-5 (A). The main requirement for this type of coupling is that the short-circuit current gain is greater than unity. Thus, the series connection in the case of the junction type may only be used in the grounded emitter connection.

Parallel-tuned resonant-coupling circuits are applicable in i-f strips, particularly when junction transistors are used. If point-contact transistors are used, special care is required to avoid oscillation due to the inherent instability of these types when short circuited. Several types of parallel-tuned coupling circuits may be employed. Figure 7-5 (B) illustrates one such possible circuit with the input of the coupled stage directly connected into the resonant circuit of the first stage. This direct coupling can also be used if the inductor and capacitor are interchanged. Figure 7-5 (C) illustrates another coupling arrangement with the input of the second stage connected to the junction of the two capacitors in the resonant output tank of the first stage. In this case, the capacitors can be used for matching the impedances between the stages. This coupling arrangement can be made inductive by reversing the reactive elements and connecting the input of the second stage into the tank inductance. This arrangement requires that a capacitor be inserted in the input lead of the second stage. This capacitor blocks the d-c bias and also helps to avoid the excessive loading of the tank due to the input circuit of the second stage. An alternate method is to couple the second stage to the tank inductively. If the inductive coupling is also tuned in the second stage, the circuit becomes the double-tuned coupling network illustrated in Fig. 7-5 (D). The center tap in the inductance of the secondary circuit provides for the proper impedance match between the stages.

Neutralization. The close coupling between the input and the output circuits of the transistor causes the resonant frequency of the coupling circuit to be particularly sensitive to variations in the input and output impedances. In general, the load impedance has a greater effect on the input impedance than the generator impedance has on the output impedance. For this reason, the best procedure to follow in aligning an i-f strip is to start with the last stage and work toward the first.

To avoid the critical tuning problem, the stage may be neutralized. This allows each resonant coupling circuit to be independently adjusted without introducing any detuning on or by the remaining stages. One form of neutralization is illustrated in Fig. 7-6 (A). For reasons of clarity, only the a-c circuit is shown. Neutralization is accomplished by balancing the resistor R_B and the equivalent impedance Z_c of R_C and C against the transistor base resistor r_b and the equivalent impedance of the collector arm z_c composed of r_c and C_c. The balancing conditions are more clearly illustrated in the equivalent circuit of the neutralized circuit, as shown in Fig. 7-6 (B). The circuit is drawn in the form of a conventional bridge. The bridge is balanced when $\dfrac{r_b}{R_B} = \dfrac{z_c}{Z_c}$. Under this condition there is no interaction between the input and output

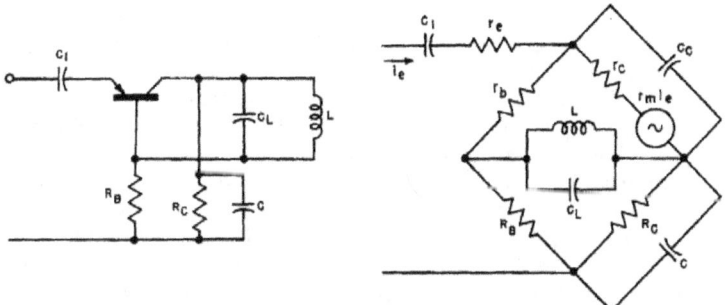

Fig. 7-6. (A) Neutralized i-f amplifier. (B) Equivalent circuit of neutralized i-f amplifier.

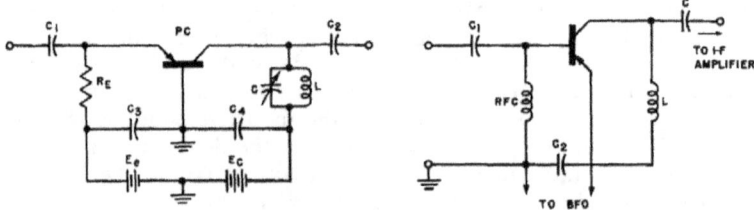

Fig. 7-7. Typical transistor r-f amplifier.　　　Fig. 7-8. Junction transistor mixer circuit.

circuits. Then, when the stage is neutralized, the output impedance is independent of R_g and the input impedance is independent of R_L.

In practical circuits the neutralization network design can be simplified by omitting the capacitor C if a point-contact transistor is used, or by eliminating R_c if a junction transistor is used. This changes the balance equations to $\dfrac{r_b}{R_B} = \dfrac{r_c}{R_C}$ for the point-contact types, and

$\dfrac{r_b}{R_B} = \dfrac{C}{C_c}$ for the junction types. These simplifications are possible at the intermediate frequencies because feedback is governed primarily by r_c in the point-contact transistor, and by C_c in the junction transistor. The network components are not very critical. Values within a 5% tolerance range are generally satisfactory.

Notice that the lower output terminal is connected to ground through R_B. This makes it important for the value of R_B to be small in order to avoid introducing too much noise through R_B into the output circuit. For satisfactory operation, the value of R_B should not be larger than the base resistance. This fixes the value of C in the range of C_c, and R_c in the range of r_c. The loss in gain due to the neutralizing network will be less than 10% of the total gain in a properly designed circuit.

R-F Amplifiers. Transistor r-f amplifier circuits, like their counterpart vacuum-tube circuit types, are most often used for improving the gain, over-all signal-to-noise ratio, or selectivity characteristic of a multistage circuit. Figure 7-7 illustrates a typical transistor r-f amplifier circuit. The design is basically the same as that of an i-f amplifier. The chief problem is the selection of a transistor having a sufficiently high α cutoff. The power gain of a r-f amplifier is inversely proportional to the frequency. In a typical case a transistor having a maximum gain of 40 db at 10 mc will have a maximum gain of 20 db at 40 mc. There are two critical parameters, the emitter bias and the base resistance. The base resistance is determined by the physical construction of the transistor and, therefore, low base-resistance transistors, designed specifically for high frequency applications, should be used. The importance of emitter bias was considered in the analysis of oscillator circuits. The bias should be selected to be far enough away from the unstable region of the characteristics to avoid oscillation, and yet not so far away that the gain is very low. Special care must be taken to avoid introducing stray capacitance into the emitter input circuit. These reactances tend to lower the input emitter impedance, and thereby decrease circuit stability.

Limiters. Limiter circuits can be designed using transistors and germanium diodes. These circuits operate much like vacuum-tube limiters. In the grounded base connection, the input circuit acts like a diode when the emitter electrode is biased slightly in the forward direction. When the value of the input signal exceeds that of the emitter bias, the signal is rectified by the diode action of the input circuit. The resulting self-bias tends to keep the maximum emitter current constant. Since the collector current is proportional to the emitter current, the output signal is maintained at a constant level over a large range of input signal values. The input rectification action is considerably improved when the circuit is shunted by a junction diode. The diode performs two important jobs. It clips large positive input pulses, and prevents the coupling capacitor from charging on extraneous noise pulses. For optimum operation, the output resistance is matched to the load, and the generator impedance is kept as low as possible.

Mixers

The operation of the transistor in mixer circuits depends upon the rectification and non-linearity of the emitter circuit when it is biased slightly in the forward direction. Figure 7-8 illustrates one basic arrangement of a transistor mixer circuit employing a junction transistor. This circuit takes advantage of the relatively high gain of the grounded emitter connection by injecting the BFO signal into the common emitter lead. The junction transistor works well in mixer stages despite its relatively low α cutoff. This is possible because only the i-f frequency must lie within the useful frequency range of the transistor. The point-contact type also works satisfactorily in transistor mixer circuit. Its utility is limited to some degree by its relatively high noise figure and low gain.

Fig. 7-9. Transistor power supply.

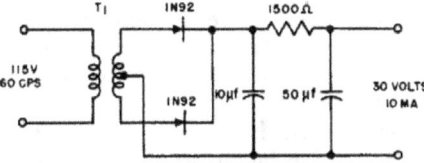

Power Supplies. As the number of applications for transistors increase, many new power supply systems will be required to fit in efficiently with the particular design. The power requirements of an individual circuit is very small, so small in fact, that quite often the life expectancy of the bias battery is the same as its normal shelf life. Nevertheless, in some applications it may be desirable to derive the power supply from an existing a-c source. Figure 7-9 illustrates an experimental power supply, fabricated for a particular application, where a bias of 30 volts and a drain of 10 ma were required. The circuit is a basic full-wave rectifier terminated in an R-C filter. With the values shown, the ripple is less than one percent.

Miscellaneous Transistor Characteristics and Handling Techniques

Transistor Life Expectancy. One of the outstanding features of the transistor is its practically indefinite life expectancy. Long life was originally predicted on the basis of the transistor construction and its conduction mechanics, which indicate there is nothing to wear out. Although the transistor is still very young, enough experimental data is now available to back the initial long life predictions.

The usual transistor failure occurs gradually over a long period of time and after thousands of hours of operation. The performance degradation generally shows up as an increasing saturation current. (The effect of increasing saturation current was covered in the transistor amplifier chapter.) While the various self and fixed biasing methods may be used to minimize the effects of increasing I_{co}, the system's efficiency and gain suffer. In an amplifier circuit, this factor decreases the available volume. Gradually, as the limit of the automatic biasing arrangement is reached, there is also a noticeable increase in the distortion content.

Another variation in the transistor performance characteristics is a gradually decreasing output resistance. In systems designed for an image impedance match ($R_g = r_i$ and $R_L = r_o$), this change introduces a mismatch loss. In the usual amplifier design, however, the output resistance is in the order of 20 to 50 times the load resistance. The decrease in r_o, therefore, is less serious than the increase in I_{co}. The best single maintenance check is a measurement of the current gain.

Sudden failures of transistors are not common in normal operation, although open emitter and collector junctions were not too rare in the early transistors. These defects were attributed to faulty assembly during manufacture. Present manufacturing and quality control techniques have practically eliminated open junction defects. Transistor shorts are more common since they are usually caused by overloading. When the transistor power rating is exceeded, the junction temperature rises quickly. The increased heating effect encourages diffusion of collector region impurities into the base layer, which, in time, will destroy the junction. In brief then, open circuited transistors generally result from poor production; short-circuited junctions generally mean improper circuit design.

Transistor Ruggedness. Insofar as ruggedness is concerned, the superiority of the junction transistor compared to the point-contact type can be anticipated from a comparison of the basic construction details (Chapter 2). The emitter and collector electrodes of the point-contact type depend on a force contact with the germanium surface. These catwhiskers, it will be remembered, are fastened to the main electrode conductors which are embedded in, and held by, the plastic stem. It is possible, then, to vary the contact pressure of the catwhiskers by a twisting force applied to the plastic stem. This distortion can be introduced by direct mechanical force, humidity or temperature variations.

Most of the present transistors are hermetically sealed. Sealing is important because of the ease with which an unprotected junction surface may be contaminated by water vapor. The contaminating effects are particularly noticeable so far as the value of the saturation current in an unsealed unit is concerned. In a typical case, the saturation current of a junction transistor will increase one hundred times its dry air value when the relative humidity is increased by 50%.

The transistor can withstand shock, vibration, and drop tests far beyond those of the vacuum tube. However, it is a good plan to treat the transistor with reasonable care to avoid unnecessary damage. The effect of distortion of the stem on electrode contact pressure was noted in earlier paragraphs. Any damage to the hermetic seal is, of course, serious. Transistor electrode leads are generally as flexible as those of regular carbon resistors. These leads should not be subjected to continual bending or flexing, or to pulls greater than a half-pound.

Soldering Techniques. Generally, junction transistors (Raytheon types 720, 721, 722, Germanium Products Corporation types 2517, 2520, 2525, Western Electric 1752, etc.) have long pigtail leads. These types can be soldered directly into a circuit. However, due to the temperature sensitivity of the transistor, solder connections must be made quickly. It is always a good idea to heat sink all solder connections by clamping the lead with a pair of long nose pliers connected between the soldered point and the transistor housing. This provides a shunt path for a large part of the heat introduced at the solder joint. If it is at all possible, transistors with short leads should not be soldered directly into the circuit. Several types of sockets will accommodate these short lead types. For example, the Cinch type 8749, type 8672, and regular 5-pin subminiature tube sockets will handle point-contact transistors similar to the Western Electric 1698, the General Electric G11A, etc.

Temperature Effects. The physical location of the transistor is not critical with respect to its mounting position, and since the heat gen-

erated by an individual transistor is small, many may be packed to-
gether. However, since the transistor is sensitive to the ambient temperature, hot spot locations near tubes and power resistors should be avoided.
In this regard, a word of precaution on collector dissipation ratings is
in order. The maximum collector dissipation is specified at some definite temperature (usually 25°C). This value must be *derated* if the
ambient temperature is greater than the specified rating temperature.
Usually this amounts to a 10% decrease in power dissipation for each
5°C increase in ambient temperature. As a numerical example, assume
that the maximum allowable collector dissipation for a transistor rated
at 250 mw at 25°C is required when the ambient temperature is 40°C.
The operating temperature represents an increase of $40° - 25° = 15°C$.
The power handling capacity should be derated 10% for each 5°C increase or $15/5 \times (10) = 30\%$. 30% of 250 mw is 75 mw. Thus the maximum collector dissipation at 40°C is $250 - 75 = 175$ mw.

Whenever a transistor is operated near its maximum rating, it is
always good insurance to tie it to a metal panel or chassis. This connection provides a large radiating surface which permits the collector dissipation to be maintained at higher levels. In typical cases, this procedure
increases the transistor power dissipation rating from 20 to 50%.

Transient Protection. In addition to its power handling limitations,
the transistor is susceptible to damage by excessive values of current
and voltage. It is particularly important to protect the transistor from
those transient surges which may be caused by switching or sudden
signal shifts. Transient effects are particularly predominant in oscillator, i-f, and high frequency amplifiers due to the storage capacity of
the reactive components. Limiting devices are usually incorporated into
the circuit. The series resistors in the emitter and collector arms of the
base-controlled negative-resistance oscillator are typical examples. In
more complicated circuits, transient limiting elements are usually selected on the basis of tests made on experimental breadboard models. If
a scarce or expensive transistor is involved, the equivalent passive "T",
made up of standard carbon resistors, can be substituted for this measurement. When connecting a transistor into a live circuit, the base lead
must always be connected first. In disconnecting the transistor from a
live circuit, the base lead must be removed last.

It is an easy matter to mistakenly reverse the polarities of bias supplies, particularly in complementary-symmetrical circuits. Reverse polarities will not impair the transistor as long as the maximum ratings are
not exceeded. It is always a good plan to check for proper capacitor
polarity, since almost all of the circuits require polarized types.

Hybrid Parameters

Significance and Derivation. The open-circuit parameters, r_{11}, r_{12},
r_{21}, and r_{22} are used exclusively throughout this book primarily because

they are the most familiar four-pole equivalents. Some engineers prefer the short-circuit conductance parameters g_{11}, g_{12}, g_{21}, and g_{22}. The conductance parameters serve well for the junction transistor, but do not work out too well for the point-contact type, which inherently exhibit short circuit instability.

The disadvantages in both the r and g forms suggest a combination or hybrid type of representation which will be applicable to all transistor types without requiring elaborate measuring techniques. The so-called 'h' or hybrid parameters are becoming more and more popular. Since many of the manufacturer rating sheets now specify the h parameters, it is important to be able to convert the hybrid values into the more familiar r form for use in the design and performance equations. On a four-terminal basis, the hybrid parameters are equated as:

$$e_1 = h_{11}i_1 + h_{12}e_2 \qquad Eq.\ (7\text{-}1)$$
$$i_2 = h_{21}i_1 + h_{22}e_2 \qquad Eq.\ (7\text{-}2)$$

The basic circuits for measuring the h parameters are illustrated in Fig. 7-10, which define the values of the parameters in terms of the input and output currents and voltages as follows:

$$h_{11} = \frac{e_1}{i_1} \quad \text{when } e_2 = 0 \quad \text{(output short-circuited)}$$

$$h_{12} = \frac{e_1}{e_2} \quad \text{when } i_1 = 0 \quad \text{(input open-circuited)}$$

$$h_{21} = \frac{i_2}{i_1} \quad \text{when } e_2 = 0 \quad \text{(output short-circuited)}$$

$$h_{22} = \frac{i_2}{e_2} \quad \text{when } i_1 = 0 \quad \text{(input open-circuited)}$$

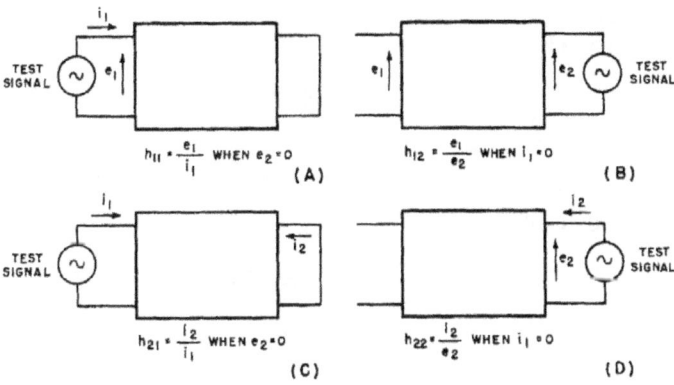

Fig. 7-10. Basic circuits for measuring four-terminal h parameters.

Notice that two of the measurements are made with the output short-circuited, and the remaining two are made with the input open-circuited. Furthermore, none of the parameters are exact equivalents, since r_{11} is a resistance (ohms), h_{22} is a conductance (mhos), h_{12} is a numeric (voltage ratio), and h_{21} is also a numeric (current ratio).

Resistance Parameters in Terms of Hybrid Parameters. The relationship between the r and h values can be determined by straightforward substitution and the simultaneous solution of equations 7-1 and 7-2, as follows:

A. $r_{11} = \dfrac{e_1}{i_1}$ when $i_2 = 0$. Under this condition

$$e_1 = h_{11}i_1 + h_{12}e_2 \qquad\qquad Eq.\ (7\text{-}1A)$$

$$0 = h_{21}i_1 + h_{22}e_2 \qquad\qquad Eq.\ (7\text{-}2A)$$

If these are solved simultaneously,

$$i_1 = \frac{e_1 h_{22}}{h_{11}h_{22} - h_{12}h_{21}}, \text{ and therefore}$$

$$r_{11} = \frac{h_{11}h_{22} - h_{12}h_{21}}{h_{22}} \qquad\qquad Eq.\ (7\text{-}3)$$

B. $r_{21} = \dfrac{e_2}{i_1}$ when $i_2 = 0$. Under this condition equation 7-2A still

applies and is solved

$$e_2 = -\left(\frac{h_{21}}{h_{22}}\right)i_1, \text{ and } r_{21} = -\left(\frac{h_{21}}{h_{22}}\right) \qquad\qquad Eq.\ (7\text{-}4)$$

C. $r_{12} = \dfrac{e_1}{i_2}$ when $i_1 = 0$. Under this condition

$$e_1 = h_{12}e_2 \qquad\qquad Eq.\ (7\text{-}1B)$$
$$i_2 = h_{22}e_2 \qquad\qquad Eq.\ (7\text{-}2B)$$

If these are solved simultaneously

$$\frac{e_1}{i_2} = \frac{h_{12}}{h_{22}} \text{ and } r_{12} = \frac{h_{12}}{h_{22}} \qquad\qquad Eq.\ (7\text{-}5)$$

D. $r_{22} = \dfrac{e_2}{i_2}$ when $i_1 = 0$. For this relationship equation 7-2B still

applies and is solved

$$e_2 = \frac{i_2}{h_{22}} \text{ and } r_{22} = \frac{1}{h_{22}} \qquad\qquad Eq.\ (7\text{-}6)$$

E. The current gain $a = \dfrac{r_{21}}{r_{22}} = \dfrac{-\left(\dfrac{h_{21}}{h_{22}}\right)}{\dfrac{1}{h_{22}}} = -h_{21}$

As a numerical example of these conversions, the manufacturer's rating sheet for the G.E. type 2N45 specifies the following typical values for the hybrid parameters:

$h_{11} = 40$ ohms, $h_{12} = 2.5 \times 10^{-4}$, $h_{21} = -.92$, $h_{22} = 1.0 \times 10^{-6}$ mhos.

Then $r_{11} = \dfrac{h_{11}h_{22} - h_{12}h_{21}}{h_{22}} = \dfrac{40 (1.0 \times 10^{-6}) - 2.5 \times 10^{-4} (-.92)}{1.0 \times 10^{-6}} = 270$ ohms

$$r_{21} = -\frac{h_{21}}{h_{22}} = \frac{-(-.92)}{1.0 \times 10^{-6}} = 920,000 \text{ ohms}$$

$$r_{12} = \frac{h_{12}}{h_{22}} = \frac{2.5 \times 10^{-4}}{1.0 \times 10^{-6}} = 250 \text{ ohms}$$

$$r_{22} = \frac{1}{h_{22}} = \frac{1}{1.0 \times 10^{-6}} = 1 \text{ megohm}$$

$$a = -h_{21} = -(-.92) = 0.92$$

Printed Circuit Techniques

One of the most promising features of the transistor is its ability to fit into the new prefabricated wiring techniques, by which the maze of hand-soldered wires normally associated with electronic equipment has been eliminated. Basically, a printed circuit starts with a metal foil bonded to one or both sides of an insulating plastic material. The metal foil may be copper (most popular) aluminum, silver, or brass. Most types of laminated plastics are suitable as the base insulator. The circuit is drawn on the foil clad laminate with an acid resistant ink. The complete assembly is then dipped into an etching solution which removes the metal not protected by ink. Holes are then drilled or punched into the assembly at appropriate points, and into these holes the various circuit components are inserted and soldered to the metal foil. If the circuit is at all complex, hand soldering is extremely tedious and difficult, and the dip soldering technique is used. In this method, components with preformed leads are inserted into the holes, either manually or by an automatic process. After fluxing, all the connections between the component leads and the circuit pattern are accomplished by a "one-shot" dip in a molten solder bath. Those portions of the circuit which must be left free of the solder are coated with a protective lacquer or masked before the solder bath. Dip soldering assures very reliable solder joints in one simple operation, and also permits a greater reduction in size by means of stacking techniques, which were previously limited by the space requirements for hand soldering operations.

Complex circuits are normally laid out on both sides of the laminated base. Connections crossovers may be made by several methods. The most common is by means of a tined eyelet. This is of particular importance in those cases where connection is made to a component

Fig. 7-11. Experimental transistor i-f amplifier.

which may be soldered and unsoldered several times during the life of the equipment. Repeated soldering at the foil will eventually cause it to lift from the plastic base.

In spite of the small cross-sectional area of the foil conductors, the current carrying capacity of the printed circuit is good, due to the relatively large surface area and the heat conduction by the base material. A 1/32-inch copper foil conductor, for example, can safely handle about five amperes. Increased temperatures caused by current overloads causes the metallic conductor to buckle and separate from the base.

One of the major advantages of the printed circuit is its uniformity from unit to unit. For example, the distributed capacitance between foil conductors is in the same order of magnitude as that of a carefully hand wired assembly. In the prefabricated type, however, the value remains constant from unit to unit because they are all produced from the same master design.

Figure 7-11 illustrates the front and back of an experimental printed circuit type of transistor i-f amplifier. The component arrangement can be seen at the left of the illustration and the printed wiring can be seen at the right. Miniature components for use with transistors are shown in Fig. 7-12. The top row of the figure shows a miniaturized transformer and three resistors. The bottom row illustrates an inductor, a capacitor, two junction transistor sockets, and two point-contact transistor sockets.

The marriage of standard and miniaturized components with the basic printed circuit is, in essence, the "autosembly" technique devised by the Signal Corps Engineering Laboratories. This method is best suited to present production facilities, since it utilized components with proven reliability. However, the recent progress in the development of printed components indicates that most of the applications of prefabri-

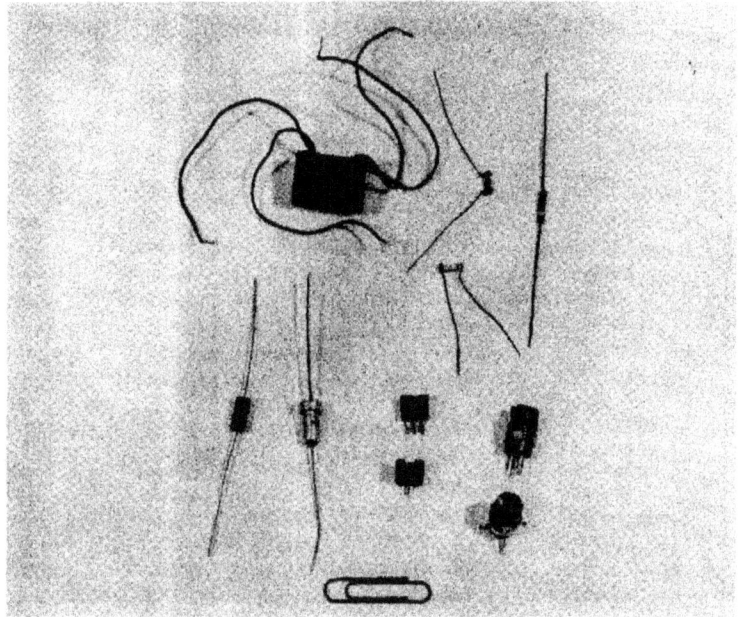

Fig. 7-12. Miniature transistor components.

cated circuits are still to come. Printed resistors having values of 10 ohms to 10 megs and which are sprayed onto an area of 1/16 of a square inch have been used successfully. Small inductance coils, having values up to 20 μh, can be etched into the printed circuit, and capacitors ranging from 10 $\mu\mu$f to .001 μf can be incorporated in the printed circuit by etching opposite sides of foil-clad glass-cloth laminates.

The transistor, because of its mechanical ruggedness and long life expectancy, is well adapted for direct assembly into printed circuit patterns. The minute heat generated by the transistor makes its future use in compact packaged equipment particularly promising. The prefabrication techniques will initially reduce the out-of-service time considerably, since complete circuits will be encapsulated in units no larger than present vacuum tubes. On the other hand, assembly repairs will require great skill and technical knowledge due to the complex arrangement of the miniaturized components.

AG	Available gain	I_e	D-c emitter current
a	Current gain	i_e	A-c emitter signal current
b	Base electrode	i_b	A-c base signal current
c	Collector electrode	i_c	A-c collector signal current
C_c	Collector junction capacitance	R_B	Base series resistor
C_e	Emitter junction capacitance	R_C	Collector series resistor
e	Emitter electrode	R_E	Emitter series resistor
E_1	Input voltage - 4-terminal network	r_i	Input resistance 4-terminal network
E_o	Output voltage - 4-terminal network	r_o	Output resistance 4-terminal network
E_{bb}	Battery supply voltage	r_1	Image matched input resistance
E_e	Emitter bias battery	r_2	Image matched output resistance
E_b	Base bias battery	r_{11}	Small signal open circuit input resistance
E_c	Collector bias battery		
e_b	A-c base signal voltage	r_{12}	Small signal open circuit reverse transfer resistance
e_c	A-c collector signal voltage		
e_e	A-c emitter signal voltage	r_{21}	Small signal open circuit forward transfer resistance
f_c	Cutoff frequency		
G	Operational gain	r_{22}	Smal signal open circuit output resistance
g_{11}	Small signal short circuit input conductance		
		r_b	Transistor equivalent base resistance
g_{12}	Small signal short circuit feedback conductance	r_c	Transistor equivalent collector resistance
		r_e	Transistor equivalent emitter resistance
g_{21}	Small signal short circuit transfer conductance	r_m	Proportionality resistance constant between emitter signal current and resulting voltage signal produced in collector arm
g_{22}	Small signal short circuit output conductance		
h_{11}	Small signal hybrid short circuit input impedance	R_g	Internal resistance of signal generator
		R_L	Load resistance
h_{12}	Small signal hybrid open circuit voltage feedback ratio	W	Base junction thickness
		VG	Voltage gain
h_{21}	Small signal hybrid short circuit forward transfer current ratio	MAG	
			Maximum available gain
h_{22}	Small signal hybrid open circuit output admittance	N-$Type$	
			Transistor semiconductor with donor type impurities (electron current carriers)
I_1	Input current - 4-terminal network		
I_2	Output current - 4-terminal network	P-$Type$	
I_{co}	Saturation current - collector current at zero emitter current		Transistor semiconductor with acceptor type impurities (hole current carriers)
I_b	D-c base current		
I_o	D-c collector current		

INDEX